ABOUT THE AUTHOR

Will McCallum has been at the heart of the anti-plastics movement for the past three years, in his role as Head of Oceans at Greenpeace UK. He regularly meets with the government and companies to implore them to help tackle the plastic crisis. He leads the global Greenpeace campaign to create the world's largest protected area in the Antarctic Ocean. He recently spent a month in Antarctica with his team, investigating whether plastic is reaching the most remote region on the planet. He is a keen long-distance runner and regularly goes sea kayaking to explore the UK coast.

You can follow Will on social media:

Twitter: @artofactivism
Instagram: @_willmccallum

WILL McCALLUM

HOW TO GIVE UP PLASTIC

SIMPLE STEPS TO LIVING CONSCIOUSLY ON OUR BLUE PLANET

PENGUIN LIFE

AN IMPRINT OF

PENGUIN BOOKS

PENGUIN LIFE

UK | USA | Canada | Ireland | Australia
India | New Zealand | South Africa

Penguin Life is part of the Penguin Random House group of companies
whose addresses can be found at global.penguinrandomhouse.com.

First published by Penguin Life 2018
This edition published 2019
003

Text design by Janette Revill
Set in 10.88/13.3 pt Sabon MT
Typeset by Jouve (UK), Milton Keynes
Printed and bound in Great Britain by Clays Ltd, Elcograf S.p.A.

A CIP catalogue record for this book is available from the British Library

ISBN: 978–0–241–38893–8

The information in this book reflects the views of the author and not the
views of the organization Greenpeace. So far as the author is aware the
information given is correct and up to date as at May 2018.

www.greenpenguin.co.uk

MIX
Paper from
responsible sources
FSC® C018179

Penguin Random House is committed to a
sustainable future for our business, our readers
and our planet. This book is made from Forest
Stewardship Council® certified paper.

To those who are struggling daily against the tide of plastic pollution across the world – all power to you and I hope this book helps a bit

CONTENTS

PREFACE

The issue of plastic pollution in the ocean and the environment has exploded into the public consciousness like no other environmental issue in recent history. Tens of millions of people across the world watched *Blue Planet II*, David Attenborough's blockbuster series, gasping in shock at the scene where an albatross fed its chicks small pieces of plastic, mistaking them for food. We have all experienced that moment of walking somewhere beautiful and seeing in our path a stray piece or pile of plastic spoiling our surroundings. Scientific understanding of the impact of plastic pollution, as well as knowledge of the solutions to prevent it getting worse, is still in its relative infancy – and yet as we understand more and more about the scale of the problem the desire to act grows stronger.

In the years I have spent campaigning against plastic the most common question I am asked is, 'What can I do to help?' *How to Give Up Plastic* provides you with the knowledge you need and guides you to make informed choices about getting rid of plastic in your own life. It would be impossible to include alternatives for every plastic product within these pages, and at the current rate of research and innovation there will be many new alternatives just months after I've finished writing, so this book includes plenty of sources where you can do your own research about products not covered here. It also equips you with the facts about the issue and the campaign tools necessary to help persuade others, including friends, family, colleagues, local businesses and politicians, to join you on the journey to creating a world where plastic pollution may be a thing of the past.

For the most part when talking about plastic in this book, I am referring to single-use plastic – i.e. plastics that are used once and then thrown away, often taking centuries to break down. Items like plastic bags, straws, coffee cups and plastic packaging. I focus on these because as well as posing an increasingly large problem to the world's oceans, these items are where we as individuals can have the biggest impact, in our own homes and communities. Also because, in my opinion, they epitomize the problem with plastic. It is not that this material – cheap, flexible and in many instances life-saving when it comes to medical uses – is inherently bad. Rather, that we have developed a throwaway culture around single use that is not healthy, for society or for the oceans – and if the plastic crisis in our seas has any silver lining at all, it may be that it provides the catalyst to snap us out of this destructive pattern.

Finally, a word at the beginning about the necessity of plastic to some people's lives, whether it be due to mobility issues requiring someone to drink through a straw or that where they live the water from the tap is unsuitable to drink – there are on occasion good reasons for using single-use plastics. These exceptions to the rule mean that in our quest to get rid of plastic it is important not to immediately point the finger before first understanding individual circumstances. Circumstances that should *not* be used as an excuse by companies and governments to forgo action and innovation to find alternatives – as explained in the excerpt written by Jamie Szymkowiak, founder of disability rights organization One in Five, on p. 117. Plastic has become so pervasive that if we want to have any chance of success, giving it up has to be a journey that brings people together, no matter what their circumstances.

INTRODUCTION

'Can I grab you quickly? Take a look at this.'

Grant Oakes, our biosecurity officer on-board the *Arctic Sunrise*, Greenpeace's icebreaker vessel, grabs me from the canteen and takes me to our makeshift lab in the hold where he has set up a microscope. As he rotates the Petri dish round beneath the lens of the microscope, I focus in on the offending object – hard, bright pink with serrated edges, it is obviously not from natural sources. It seems we have found our first fragment of plastic in the pristine Antarctic waters we are sailing in. A couple more colleagues join us and we take turns to examine it. We cannot confirm whether it is plastic until next month when we will take it back to our laboratory in Exeter University for analysis, but to the untrained eye it is hard to imagine it is anything else. (A few weeks later the results revealed that we had found two fragments of plastic in waters hundreds of miles from permanent human habitation.)

The reaction in the canteen is far from surprised; if anything we expected to see these results sooner. Greenpeace ships have been testing for plastics in the ocean since the mid-1990s and in the past few years its presence in our trawls has increased more and more across every ocean the ships sail in. As is becoming common practice on each of Greenpeace's three ships, we are taking every opportunity, weather and ice permitting, to do what is known as a manta trawl – using a fine-mesh net, about a metre across at the mouth, to test for plastics in the water. From the frozen Arctic tundra to the deepest trenches in the ocean, scientists have found plastics almost everywhere they have searched, so why not down here in the Antarctic as well, at the bottom of the world, despite its lack of permanent inhabitants.

We have been sailing here for nearly two months now. We are working with scientists, journalists and celebrities to raise awareness about the need to protect this vast wilderness. It is a landscape like none other I have experienced – much of the time clouded in deep fog that occasionally lifts to reveal dramatic peaks and enormous glaciers cascading down into the water. The main topic of conversation on-board is the unbelievable abundance of wildlife around us at all times. Fix your gaze on a stretch of water for long enough and you are almost guaranteed to spot the fin of a humpback whale dawdling past or a small group of penguins porpoising out of the water in between the contorted icebergs that surround us. It is sobering to think that even these icy waters, teeming with wildlife wholly unconcerned with us humans, are beginning to be polluted by plastics being produced halfway across the world.

It doesn't take a trip all the way down to the Antarctic to come to this grim conclusion. Everybody I speak to about the issue has had the experience of plastic encroaching on a beloved landscape. It is nearly impossible to visit our favourite beaches or walk along a river without spotting some floating plastic debris drifting menacingly out to sea. The problem of plastic pollution is one close to the heart of so many people because it is affecting all of us, every day. From the tabloids plastering it on the front pages to politicians having lengthy debates in the Houses of Parliament, from ordinary households attempting to go plastic free to celebrity endorsements of products that purport to be better for the environment, everyone is talking about the need to find a solution to the tide of plastic flowing into our oceans.

People across the world have woken up to the ridiculousness of the situation we are in: we managed to create a material and use it at unbelievable scale with no plan for how to deal with it afterwards. Single-use plastic cutlery, plastic bags and plastic-lined coffee cups have become central to our lives – used once for a matter of minutes, they will not break down for hundreds of years. It is untenable to carry on like this: we are consigning future generations to a world in which plastic might outweigh fish in the ocean by 2050. This mind-boggling statistic, together with our shared anger at over-packaging and useless plastic products, is galvanizing a global movement prepared to go beyond just talking about the problem and to actually start doing something about it.

This book is for those who want to act now but don't know where to begin. In the face of a problem on this scale it can be hard to work out what your role is; whether you can actually make a difference. I don't claim to have all the answers, far from it, but having spent a few years campaigning to reduce plastics, talking to people about their experiences, negotiating with companies and politicians about what can be done, I've compiled this useful guide to help you play a part in ending ocean plastics. From the kitchen cupboard to the boardrooms of multinational companies, the movement to end plastic pollution needs everyone to come on-board and do what they can – at home, where they work and in their community.

If you take one message from the book it is this: that the problem of plastic pollution is one that affects us all, and therefore one for which we all share responsibility as individuals but also, more importantly, collectively. As individuals we can change our behaviour, limit our use and help reduce, even by a little bit, the amount of plastic out there. Working together we can achieve much, much more. Amplifying your actions by talking about them with your friends, colleagues and on social media, you can have so much more impact than only working behind closed doors; and joining forces with others in your community to send the message loud and clear to those with more power in politics and business is perhaps the best opportunity we have to get to a world without plastic pollution.

The problem of plastic pollution is one that affects us all, and therefore one for which we all share responsibility as individuals but also, more importantly, collectively.

MY TOP FIVE STEPS FOR GETTING RID OF PLASTIC

Just in case you get no further than this introduction – you lose the book or don't have time – then in the spirit of being a useful guide for anyone, no matter who they are and what their circumstances are, here are my top five steps for getting rid of plastic, right at the very beginning.

1 **Go on a plastic-free shopping spree.** Who would have thought in a book about reducing the amount of waste we produce the top advice for getting rid of plastic was to go and buy a few things? Essential items for a plastic-free life include: a nice water bottle, a reusable coffee cup, a tote bag (or even just a backpack) for your shopping, a lunch box and some kitchen storage containers.

2 **Go on a plastic-free purge.** Start in your bathroom, work your way to the bedroom and then into the kitchen. Have a look at ingredients lists on the back of your cosmetic products to check there aren't any microbeads; empty your cupboards of single-use plastic straws and cutlery. Don't know what to do with it all? You could always send it back to whoever you bought it from with a message that, in your household, single-use plastic is no longer welcome.

3 **Do some plastic-free preaching.** All of us are way more likely to take advice if it comes from our friends and family, rather than just reading about it in a book or watching it on the television. Pass on handy tips to your friends and neighbours (you could even give them a copy of this book). Spread the good news that a plastic-free life is easier than they think, and every little bit helps.

4 **Make some plastic-free plans.** It's true that getting rid of plastic takes a bit of planning. Use a rainy day to sit down and work out which shops near you already use less plastic. Do you have a local greengrocer that lets you pack your fruit and veg however you want? If there are only fast-food outlets near your place of work, spend some time making food for a week of packed lunches. Start thinking about your plastic-free routine and write it down in your diary.

5 **Start your own plastic-free campaign.** Go out in your neighbourhood and see what businesses are using too much plastic, and which ones are the local champions. Talk to business owners about what they could be doing to use less plastic. Why do they only use plastic cutlery and single-use coffee cups? Have they ever thought about using cardboard trays instead of styrofoam? Ask your friends to join you in asking these businesses to change their ways – after all, the customer is always right!

1

A SHORT HISTORY OF FIGHTING PLASTIC

Ban the Bead

A few years ago no one could have imagined that the world might be up in arms about tiny balls of plastic. Most people, myself included, had never heard of microbeads – minuscule fragments of plastic under 5 mm in diameter that had sneakily been added to many household products, designed specifically to get washed down the drain, with no thought as to where they would end up. Then in December 2013 a new research paper came out, revealing the extent of plastic pollution in the Great Lakes of Canada and the United States of America. Lake Ontario, the smallest of the lakes, was estimated to have as many as 1.1 million microbeads per square kilometre.

A campaign quickly kicked off, and within two years the United States Congress had passed a law to ban microbeads from many (though unfortunately not all) products. Former president Barack Obama had capitalized on the fact that the Great Lakes not only spanned two countries, but were also one of the United States' most iconic holiday destinations, a centre for business and home to over a fifth of the world's supply of fresh water. Protecting them from pollution is an issue that bridges the political divide. News of the ban reached the UK and though some of us working in the world of ocean conservation were dimly aware of the microbeads issue, it was only at this point that we considered making it a New Year's resolution to get an equivalent ban in the UK. After all, if Obama could ban these tiny balls of plastic, why couldn't our government do the same?

Of course, this was not the first major move to prevent plastic pollution, nor the most impressive. After devastating floods at the turn of the century were found to have been exacerbated by plastic bags clogging up the drainage system, in 2002 Bangladesh became the first country in the world to ban them (though plastic is so persistent, plastic bags continue to cause big problems there). Campaigners like Annie Leonard, who founded The Story of Stuff Project, had already made hugely successful viral online videos looking at how ridiculous single-use plastic is. In the UK, in autumn 2013, Deputy Prime Minister Nick Clegg announced a 5p charge on plastic bags in larger stores after a successful campaign by the Marine Conservation Society and others – a charge set to be extended to smaller retailers after being found to be responsible for an 85 per cent reduction in plastic bag use. Across the world, from sub-Saharan Africa to San Francisco, a growing movement against plastics was already gaining power.

In January 2016 we at Greenpeace UK launched a petition against microbeads, quickly forming a coalition with other organizations working on this issue – the Marine Conservation Society, Fauna & Flora International and the Environmental Investigation Agency. The response was like nothing we could have predicted. Very soon hundreds of thousands of people had signed our petition to ban the bead, newspapers like the *Daily Mail* had plastered our campaign across their front page and celebrities were lining up to join the movement. Pent-up anger over plastic pollution evolved into public outrage that these microbeads had made their way into the market; customers felt fooled, they had had no idea that their face wash was responsible for thousands of microbeads leaking out to the ocean.

As a campaigner, it was a gift – a simple ban presented an easy solution and what's more it had popular support. All we had to do was channel the outrage in the right direction, towards the minister who could make change happen; work with our coalition partners to continue to build the evidence for why a ban was needed and what a ban might look like in law; and encourage companies to keep up momentum by voluntarily committing to stop stocking microbeads in the meantime. However, it soon became clear that this was just the tip of the iceberg when it came to people's frustration about plastic and, indeed, when it came to the scale of the problem itself. Each morning I was arriving at work to an inbox filled with questions and suggestions about what else we could be doing to get rid of plastic.

Making Plastic Bottles a Thing of the Past

The door was open, then, to expanding the campaign further. We began our hunt for what to focus on next with two questions. First of all, where is all this plastic in the ocean coming from? Secondly, where do we think Greenpeace can have the biggest impact in stopping it getting there? As an organization famous for taking action to protect the environment, Greenpeace is often the first port of call for people wanting to do something about the environmental destruction they see around them. We had the opportunity to lead by example and help shape the fight against plastic. In searching for answers we spoke to everyone – from scientists to CEOs, from Greenpeace supporters to journalists. Very quickly it became obvious that despite the enormous scale of the problem, as a fairly new issue relatively little research had been published compared to other environmental problems. We wrote our own literature review about the problem of microplastics in seafood, which found that in the previous two years more papers had been published on the issue than in the three preceding decades combined. It also became obvious that launching a campaign against plastic pollution would be no short-term project; if we genuinely wanted to change things we could be campaigning for many years to come.

With such a blank slate, where to begin? Every year Ocean Conservancy publish a report following their international coastal clean-up, an annual event involving more than half a million people in over a hundred countries collecting and recording items of rubbish on their local beaches. The report reveals a list of the most-found plastic items on beaches and observed at sea. Every year the results remain roughly the same: coming in at the top are cigarette butts, accounting for more than a fifth of all items collected year on year, but both plastic bottles and plastic bottle caps consistently make it into the top five – combined, they would take the top spot. From research my colleagues had done, interviewing a wide range of people to find out what makes them tick when it comes to plastic pollution, we had found that plastic bottles were definitely something that made people hot under the collar. Intrinsically, we understand how ridiculous it is to buy a bottle of water or a fizzy drink, and then throw this perfectly good container away after a single use, but we still continue to use 35 million of them every day just in Britain.

Less than half of the 13 billion plastic bottles that

British people throw away every year are recycled. Coca-Cola, the world's largest producer of drinks sold in plastic bottles, estimates that it produces over 120 billion per year – if you laid them down nose to tail, that's enough bottles to wrap around the circumference of the earth nearly 700 times. It is no wonder then that so many end up in our rivers, on our beaches and, eventually, in the ocean. If Greenpeace wanted to have impact, plastic bottles were an obvious place to start.

Less than half of the 35 million plastic bottles that British people throw away every day are recycled.

What to do about this ubiquitous product? The first goal had to be reducing the number of them out there. We simply cannot continue to produce the amount of plastic we do currently – no waste or recycling system in the world is equipped to deal with the pure volume of rubbish we are creating. Companies producing bottles in these quantities need to start looking at pilot schemes that move us away from single-use containers: drinks fountains, refillable and reusable bottles, for example. There may also be some room to use alternative materials, but because any material used in this quantity is likely to come with some negative impacts, the quest to find the best alternative material to plastic should not come at the expense of alternative delivery systems.

After reduction, and bearing in mind that it could be a long while until plastic bottles are a thing of the past, we turned to what to do in the immediate. We joined the Campaign to Protect Rural England campaign for the government to introduce a deposit return scheme, mirroring the well-known milk bottle scheme in which you pay a small deposit extra for each bottle you buy and you get your money back when you return them. Schemes like this have been successful in recapturing over 90 per cent of plastic bottles in Germany and Norway. Alongside this we also campaign for companies to commit to increasing the amount of recycled content in the bottles they produce. By creating a demand for the recycled material, there is an additional incentive to recapture the bottles through measures such as a deposit return scheme, rather than allowing them to be lost into the environment or end up in landfill.

The campaign is working – perhaps not quick

enough for my liking, but we are seeing change happen before our eyes. The Scottish Government has announced its intention to implement a deposit return scheme and in March 2018 the UK's minister for the environment, Michael Gove, announced his intention to implement such a scheme nationwide. Coca-Cola have made a public commitment to recapture every single one of the billions of bottles they produce each year (quite how, remains to be seen). The commitments may still be no more than promises on paper so far, but they are a sign that across the board companies and politicians are waking up to the need to act now to get rid of plastic.

It is quite rare in environmental campaigning to feel as though you are on the winning side. To be honest, when I go kayaking on the west coast of Scotland and see the beaches we camp next to covered in plastic waste it can be easy to forget that, as campaigns go, this one is in fact doing okay – the movement is growing fast. I am used to working on campaigns where change is incremental and public interest is comparatively low. Working at Greenpeace, you are used to being on the outside trying to be heard by those in power. It is an unusual experience to have politicians, journalists and company executives clamouring to hear what we have to say on a subject. The call to end ocean plastics is coming from so many directions that although we may disagree at times on the best way to go about it, it feels as though we are riding a wave where tectonic changes are possible.

Getting to the Root of the Problem

I made the decision to start campaigning on plastic pollution because there was a clear Greenpeace-shaped hole. Despite years of campaigning for incremental policy change by some organizations, the vast majority of campaigns and communications in the public sphere about plastics seemed to be focused on making ordinary people and consumers feel guilty about how much plastic they were consuming and failing to recycle. Nowhere did it seem to be acknowledged that actually, despite the best of intentions, it is nigh on impossible to give up plastic altogether and to not be complicit in how much is out there. Although it's clear there is so much people can do as individuals to make a change, it's also clear that producers of plastic packaging are making far too much of it, with no plan for what to do after it is used once. Politicians simply aren't going far enough in making producers take responsibility. It is not your fault if your local recycling scheme is not equipped to deal with the volume or type of plastic found in your local supermarket – and so to place the full responsibility on individuals to deal with it cannot be fair.

This discrepancy between who in our society seemed to be being held accountable for the problem was what brought me to the conclusion that Greenpeace needed to get involved. We needed a campaign to make sure that everyone – including companies and politicians – feels the pressure to do their bit to get rid of plastics, even if it means some quite radical changes. Sharing the responsibility for plastics in the ocean between individuals, the state and companies – as well as across borders – is the only chance we have at collectively finding the solutions we need to resolve the problem. That is why this book goes beyond just acting in your own home, as important as this is, and instead I'm sharing with you everything I and my colleagues have learnt about dealing with corporations and governments in the hope that you too can amplify the message within your own community about the need to give up plastic.

To place the full responsibility on individuals cannot be fair.

Since the beginning, I've shared this journey with many colleagues both inside of Greenpeace and outside. One of those colleagues is Luke Massey, whose campaigning nous and storytelling has defined the way the issue is perceived in the public consciousness. Here are his thoughts on plastics:

Who are you?

I'm Luke Massey and I am a press and communications officer at Greenpeace working on oceans issues.

Why do you care about plastic so much?

The terrible scale and impact of plastic pollution on wildlife is heart-wrenching. But to me, how we as a species come to terms with the plastic pollution crisis speaks to an even bigger question. That is: how do we move from a throwaway culture globally, to one where we minimize our footprint on this planet? There's not only a demonstrable good in stopping the flow of plastic into our environment, but in dealing with this, we are forced to reimagine our relationship with the things we produce and consume. If we learn the right lessons, the results could be transformative.

What can people do to help?

You can do more than you think. Most of the major changes I've seen in the last few years have come from people talking to other people in their community about this issue – talking to businesses, writing letters to the local newspaper or politician. Those conversations are what has made this issue take off in such a huge way.

What's the worst example of plastic pollution you've seen?

The saddest example I've seen was on a remote penguin colony in Chilean Patagonia. This island is in the middle of nowhere, and penguins were in nesting season. They dig their nests underground and the chicks stay in the sheltered warmth until they're a little older. Next to one nest I saw a male returning with a mouthful of plastic packaging from the sea and delivering it to the nest. It was utterly depressing.

What's the best solution to reducing plastic you've come across?

While showy devices which claim to clean up plastic waste in the ocean get a lot of attention, the solution has to come at the source. So, while it's a startlingly mundane answer, it's absolutely critical: taxing producers of throwaway plastic. Seeing governments move towards a 'polluter pays' approach to single-use plastic is really encouraging. To tackle the problem at source, it has to be in the interests of major plastics producers to innovate away from a throwaway business model.

Are there any changes you've made in your life to reduce plastic?

As a coffee addict, I think the biggest reduction in my plastic footprint has been getting a reusable cup. I probably buy one or two coffees a day, so the sheer scale of personal reduction over the course of a year is pretty huge. When I was first bringing it into coffee shops, it always seemed like a bit of an event for them. Now it's commonplace and most cafes offer discounts for using a reusable cup.

What makes you most annoyed when it comes to plastic?

Businesses passing on the blame. For years companies have been profiting from pumping out tons of single-use plastic on to the market and taking zero responsibility for the end life of their products. Instead, companies have blamed the public for littering. I'm sick of business models which profit from plastic but let everyone else sort out the mess.

Do you have any top tips for getting rid of plastic?

Use less. We all need to dramatically reduce the amount of throwaway plastic we consume.

Reuse. Get a reusable water bottle and refill it. Take a reusable coffee cup to cafes. Take a bag to the shop.

Recycle. It goes without saying that we should recycle whatever we can.

Talk to people. Talk to friends, local businesses. Ask why shops are selling unnecessary plastic.

What's the most impressive effort to reduce plastic you've seen – from an individual or a company?

Back in 2016 a guy in New York called Rob Greenfield decided to wear every single item of trash he produced for a month. Bags, containers, coffee cups, plastic bottles, the whole shebang. Ballooning into a trash monster, he took to the streets of New York to engage people on the issue of consumption and waste. It was a talking point rather than an initiative to reduce waste, but it helped to spark a huge conversation in the media about the issue of consumption and plastic pollution.

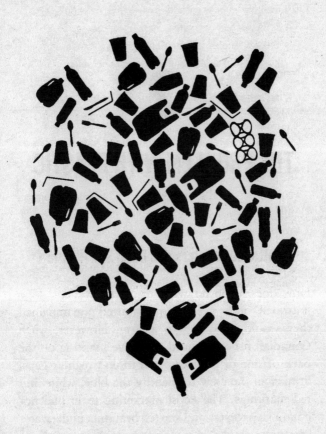

2

THE PROBLEM WITH PLASTIC

Tattooed Lobsters and Deep Sea Plastic

How does a lobster wind up with the Pepsi logo 'tattooed' on its shell? A question no one imagined they would be trying to answer. However, when Canadian fishers noticed a strange marking on the back of one of their catch, it took a regular Pepsi drinker in the crew to identify the blue, white and red markings. The Pepsi marketing team had not gone a step too far and started branding underwater creatures; rather, it was just another example of how we humans are making our mark on the ocean through the rubbish we throw away. The finding made global headlines, a shock story about how far things had come – but amongst my colleagues and those familiar with the issue it was not met with surprise, but with sad resignation.

In summer 2017, my team led an expedition around the coast of Scotland on a Greenpeace ship called the *Beluga II*, a small yacht that can sleep twelve people at an absolute maximum. The purpose of the expedition was to document the presence of plastics in the feeding grounds of some of the UK's most iconic wildlife like puffins and basking sharks. Basking sharks are the world's second largest fish; enormous, mysterious sharks described by the legendary Scottish poet Norman MacCaig as a 'room-sized monster with a matchbox brain'. It is a source of great frustration that I have yet to see one. Kayaking through their feeding grounds on several occasions at the right time of year, longingly looking down in the hope of seeing one pass beneath me in the depths, they continue to elude me.

Regularly growing to over ten metres long, they dawdle across the world's oceans feeding on tiny plankton. Their mouths can be over a metre wide, permanently open in search of food, which specially formed bones in their mouth filter out of the water. The width of their mouth is equivalent in size to the fine-mesh nets we trawl alongside the Greenpeace ships to test for microplastics; these prehistoric beasts must now be consuming vast amounts of our modern-day waste along with their dinner. Over the course of a couple of months the team undertook nearly fifty of these trawls, finding microplastics in two-thirds of the samples.

What are Microplastics?

Microplastics are tiny pieces of plastic, sometimes defined as being less than 5 mm long. Some microplastics – such as microbeads – are designed that small. However, many microplastics are fragments of a larger body of plastic such as a plastic bag or bottle that is gradually breaking down into smaller and smaller pieces.

Throughout the expedition, a team following the ship's route by land undertook beach cleans and detailed surveys at every opportunity, often working with local schools and community groups to pick up and document the plastic waste being washed up ashore. Every single beach, some of them amongst the most beautiful and remote of the British Isles, was littered with plastic: wet wipes, bottles, bags and countless more casually discarded items. Listening to the endurance swimmer and UN Patron of the Oceans Lewis Pugh speak to a group of politicians, I was moved to hear about a similar experience he had on a beach clean he led on Barentsøya (otherwise known as Barents Island), an uninhabited island in the Svalbard Archipelago, far north in the high Arctic. Despite never being home to human residents, he and the marine biologists accompanying him found the island's beach covered in bits of plastic. Some of it, such as old fishing gear, may have drifted on to the beach from relatively nearby, but the majority had been carried by the ocean currents, possibly travelling several thousand miles. Gathering an enormous sack of plastic within less than an hour was heart-breaking enough, but it was devastating to see, several days later after some rough weather, that the beach was just as covered as it had been before. As he wrote on the matter: 'Barents Island is for polar bears, not plastic.'[1]

So many remote regions of the world are in a similar state, for example the Mariana Trench in the western Pacific Ocean. Measuring up to 11 km below the surface of the ocean, it is the deepest place on earth, and one of the most mysterious on the planet. Scientists from Newcastle University discovered that every single sample they had taken from the deepest point of the trench contained microplastics. Amphipods, tiny shrimp-like bottom feeders, which could never even have seen sunlight before, had ingested tiny fragments of plastic. The highest density of plastic ever to have been documented was on an uninhabited coral atoll called Henderson Island in the South Pacific. Scientists studying the island estimated there were over 38 million pieces of plastic, with items from Germany, Canada and other distant places recorded. The sad

1. http://lewispugh.com/no-place-to-hide-from-plastic/

conclusion is that no matter how hard we work to get rid of plastic now and in the future, the impact of our actions is so vast that it will be visible across the globe for generations to come.

The Impact on Wildlife

These spectacular, far-flung places littered with our waste are, of course, more than just landscapes. They are safe havens to fascinating creatures and complex ecosystems, home to species that have evolved to the conditions particular to that place – conditions which are already increasingly threatened by a changing climate and which do not need the additional burden of plastic pollution to deal with. Research on the extent to which some species are impacted by plastic is still emerging; however, what is clear is that few ocean creatures have remained untouched. In 2015 a group of Australian scientists published a ground-breaking study in the *Proceedings of the National Academy of Sciences* which estimated that over 90 per cent of seabirds were likely to have plastic in their guts. It's possible on reading this statistic that the famous image taken by the photographer Chris Jordan comes to mind. He made headlines with his picture of a dead albatross chick in the North Pacific, its decaying body revealing a stomach filled with fragments of plastic, killing it before it had the chance to flee the nest. For me it certainly remains one of the most iconic images relating to plastic pollution ever taken.

I am a birdwatcher, not an obsessive – my skills at identification are nothing to write home about – but sitting for an hour with a pair of binoculars watching these incredible beasts swoop and glide above or peck their way across the seashore is one of the most rewarding ways I know to experience the natural world. Whether it's the daredevil gannets diving into the water around me at speeds of up to 100 km/h, the delicate and occasionally vicious Arctic terns that spend their lives travelling back and forth between the polar regions or the mighty albatrosses that ride amongst the roaring skyscraper-size waves of the Southern Ocean, so big they can only take off in flight when the winds start blowing at near gale force, seabirds are some of the hardiest creatures on earth. As the ocean they depend on changes quickly, their food sources are increasingly scarce and many colonies are struggling to adapt. Studies show that seabird populations have declined by as much as 70 per cent in the past few decades. The last thing these beautiful wild things need is additional problems caused by floating bits of plastic filling up their stomachs.

On one Greenpeace expedition, my colleague, the wildlife photographer Will Rose, spent three days camped out on the remote Shiant Isles, documenting the puffin colony there. Puffins are hardy sea parrots with oversized beaks that can bob around in some of the roughest seas on the planet, completely unfazed, for months at a time before returning to the same network of hillside burrows to nest underground. Like other seabirds, they are coming under increased pressure from climate change. Even on these fairy-tale-like islands nestled between the Outer and Inner Hebrides off the west coast of Scotland, he captured a haunting image – a puffin, proudly perched on a rock its ancestors have called home for potentially thousands of years, carrying in its beak a dainty strip of light-green plastic.

Over 90 per cent of seabirds are likely to have plastic in their guts.

Seabirds are not the only victims, of course. Whenever I arrive in a new city for work one of the first things I do is go out running – years of living aboard a narrowboat means I head for the city's waterways, trying to explore by foot its network of ponds, lakes, rivers and canals. No matter where in the world I go, I come across a familiar sight. Whether it's the moorhens on London's Regent's Canal, the night herons beside Oakland's Lake Merritt, the gulls over the River Elbe in Hamburg or the bulbuls in Taipei's City Park – spend long enough watching and you are bound to come across the all too familiar sight of birds making their nests with plastic or pecking their way through it in search of food. The plastic that litters these inshore waterways is of course due to be taken by the current out into the ocean where it is not just birds that are at risk.

Whether it's turtles mistaking plastic bags for jellyfish or sperm whales mixing up our rubbish with the squid they hunt for in the deep sea, products that are taking centuries to break down in the ocean are posing a major threat to ocean creatures. One of the more obvious ways it can cause problems is through entanglement. A 2014 report prepared for the US Government found that young sea creatures, in particular seals, were prone to getting tangled up in debris, and that over 200 different species had been recorded in US waters suffering from entanglement (although they also noted this was likely to be a conservative estimate). Ingestion and entanglement unfortunately aren't even the only ways in which plastic can hurt wildlife. Plastic is being eaten by everything – from the tiniest plankton to the great whales, it is entering the food chain at every level.

Toxins

It is a well-known phenomenon that the higher up the food chain you get, the greater the chance of toxins accumulating and becoming more problematic. This process is known as bioaccumulation. One of the more famous examples is that of mercury, regularly seen at extremely high levels in tuna and other predator fish like swordfish where it accumulates in the muscle tissue. Sitting at the top of the food chain, we humans are often the ultimate destination for this toxic matter (although fortunately food standards do a reasonable job of keeping the worst off our plates). Before it gets to us humans, however, as might be expected, bioaccumulation can cause a range of problems for the predators themselves such as making them sick or preventing them from reproducing.

Another famous toxin example is that of poly-chlorinated biphenyls (PCBs), a group of chemical compounds used widely since the 1930s in some products (for example flame-retardant coatings, fluorescent lighting and more) until countries began to ban them in the 1970s, culminating in a global ban in 2002. These polluting compounds can leak into the environment through industrial processes, where in a similar process they are seen to accumulate in the blubber of marine mammals like whales and seals. Once PCBs have accumulated beyond 'safe' levels they can cause many problems such as lowering the immune system, making these animals more susceptible to parasites and preventing them from reproducing. In a particularly sad phenomenon, marine mammals will metabolize their blubber and the toxins in it and end up feeding them to their young through their milk.

How does this relate to plastic pollution? Ocean plastics, as well as being made up of their own set of chemical compounds that could prove problematic, can act as sponges. Researchers from San Diego State University found that after it enters the ocean, plastic can continue soaking up other toxins in the water, including PCBs. This means that before being eaten by an unwitting fish or ingested by a mussel or oyster, its toxicity has increased and the bioaccumulation process is therefore exacerbated. Research into this phenomenon is still in the early stages, but these initial results are worrying. Research into how this accumulating toxicity in seafood impacts on human health is also in its relative infancy, and it's too early to draw any clear conclusions. Researchers from Vienna shocked the world when they published research results showing plastic was found in every sample of human poo they tested. However, this must become a priority for government public health bodies given the growing prevalence of plastics on our dinner plates.

How Does Plastic Get into the Environment?

Having looked at how far plastic has spread and the ways in which it is impacting on wildlife, it is time to look at how it actually gets into the environment in the first place.

Here are some answers to the three big questions I normally get asked. First, how much plastic is already in the ocean (and can't we clean it up)? Working out exactly how much plastic is already in the ocean is a tricky business. Different factors make such estimates difficult: for example, which kinds of plastic float and are therefore more likely to be found on beaches, compared to ones which sink out of sight down to the seabed; the invisibility of so much of the plastic in the ocean, like microfibres and microplastics, which can be impossible to see with the naked eye; and perhaps most importantly, the scale of the ocean itself! Over two-thirds of our blue planet is covered by the ocean, and we have observed only a minute fraction of the seabed, making any comprehensive survey of how much plastic is out there remarkably difficult.

Despite the difficulties, Ocean Conservancy working with the McKinsey Center for Business and Environment has estimated that there is already 150 million tonnes of plastic in the ocean. This is a worrying enough figure; however, the Ellen MacArthur Foundation estimates that at the rate we are increasing production, plastic could even outweigh fish in the ocean by 2050.

It's hard to imagine any technological or human effort managing to clean up 150 million tonnes of plastic. That's the equivalent weight to 300 of the tallest tower in the world, the Burj Khalifa, but spread across every ocean in the world, from the sea surface to the deepest trench. In short, it probably can't be done, which is why efforts to clean up, though laudable and necessary in particular locations where plastic is causing major environmental or infrastructure problems, can only have limited success. The global movement to reduce plastic needs to focus on getting rid of it at source as the only way to prevent it from getting into the ocean in the first place. Of course, we need to pick up plastic on our beaches, and governments and companies need to sponsor efforts to do mass clean-ups, but unless we start reducing the amount of plastic overall, we'll only end up in an endless cycle of having to repeat these clean-ups.

It is estimated that there is already 150 million tonnes of plastic in the ocean.

Second, how much plastic enters the ocean every year? At current estimates between 4.8 million and 12.7 million tonnes of plastic waste enter the ocean every year. That's almost a rubbish truck every minute, and a recent report by the UK Government about the future of the ocean estimated that the amount of plastic waste entering the ocean could treble in the next ten years.

Third, where is it all coming from? This is a harder question to answer. Research shows that approximately 80 per cent of all plastic in the ocean originates on land rather than from ships at sea. It can end up in the ocean in many ways including:

- Microfibres, which can be released through washing our clothes, account for roughly a third of plastic in the ocean.
- Plastic that isn't disposed of correctly leads to litter blowing into waterways and being carried out into the ocean.
- Plastic that doesn't get recycled can end up going to landfill sites on the coast that might leak into the sea, as well as many other routes to the open ocean where currents will take it to the four corners of the globe.

The enormous quantity of plastic being produced means that even vastly improved waste and recycling infrastructure would not capture all plastic. Even a proportionally small amount of leakage (plastic that is meant to be disposed of properly but 'leaks' from the system) can have major impacts. In fact, scientists estimate that currently nearly a third of all plastic waste manages to evade the waste and recycling sector.

Roughly a third of plastic in the ocean comes from microfibres which can be released when washing our clothes.

Recycling

Plastic production has gone sky high in the past twenty years, reaching over 320 million metric tonnes in 2015, heavier than every human alive on earth today put together. What's more, this figure is set to double in the next twenty years. Ever since the invention of the plastic bag in the 1960s, the first plastic item cheaply available for mass consumption, plastic has become more and more a part of our lives and our society. It is now so widespread it can be hard to conceive of a world without it. The mass production of single-use plastics – items that are used once, for minutes, before being thrown away, where they may take centuries to break down – is an issue that has come to the forefront in the past few years. While the use of these items has seen an exponential increase, there has been relatively little development in the recycling and waste infrastructure needed to deal with it, leading to even more plastic leaking into the environment. Only 14 per cent of plastic ever produced has been collected for recycling and approximately 5 per cent has actually been recycled (rather than down-cycled).

Waste facilities vary the world over, but even in countries with more advanced infrastructure such as the Netherlands or Japan, plastic production has far exceeded their ability to cope. We are now at a cross-roads where we can choose to carry on with the uphill battle of trying to develop the waste industry to such a level that it will deal better with the increased pressure (although there will always be significant leakage into the environment) or rethink our approach to product design and move away from single use and towards a more holistic approach that considers the full life cycle of everything we make. I think the choice is an obvious one. There is no evidence that any waste or recycling system anywhere on earth is capable of processing the amount of waste that we are producing – not without significant environmental consequences, such as in the case of incineration which can release toxins into the atmosphere as well as increase carbon emissions. Of course, developing our waste infrastructure has to play a part in reducing the amount of plastic in the environment, but it is not a good enough long-term solution that we can rely on worldwide.

Trading in Waste

In addition to problems in domestic waste systems, a global waste trade transports millions of tonnes of plastic around the world every year. Countries that do not have the space, infrastructure or inclination to deal with their own waste sell it to other countries to deal with, shipping it through a complicated network of dealers halfway across the world. Even if you do dispose of your plastic responsibly, it may end up being sold and shipped abroad to either be recycled, incinerated or disposed of elsewhere if it becomes contaminated. At the end of 2017, China announced that it would no longer receive plastic waste from other countries, its waste systems no longer able or willing to deal with plastic coming from elsewhere given increased rates of domestic plastic production. European and North American countries now have to search for new destinations for their waste. One possible outcome of this is that waste from the West could end up maxing out the capacity of infrastructure in South East Asia, making it even harder for some of these countries to deal with the problems arising from their own plastic use.

This is why attempts to point the finger at other countries for letting more plastic get into the ocean should be viewed with scepticism. Although it may technically be the case, there are likely to be a variety of factors such as waste exports, lack of safe drinking water requiring plastic bottles, extreme weather events ruining infrastructure and lack of investment by companies producing the plastic in the first place that are making it near impossible for them to deal effectively with the amount of waste being produced.

One such example is the Philippines which, according to a 2015 study published in *Science*, is the third worst polluter into the oceans. However, a 2017 clean-up effort in Manila Bay coordinated by Greenpeace and the Break Free From Plastic movement collected 54,620 pieces of plastic, documenting where possible the brand of every item they collected. Out of the top five worst offending companies, three are well-known multinational corporations: Unilever, Nestlé and Procter & Gamble. These are companies that in some countries are going to great lengths to demonstrate their sustainability credentials; however, clearly they are still a major part of the environmental problem on the other side of the world.

One of the biggest issues with these fast-moving consumer goods companies is their increased production of sachets. Sachets allow for very small quantities of liquids to be packaged and so are ostensibly providing a cheaper way for people to buy household products like shampoo. Unfortunately these sachets are generally made using a non-recyclable combination of a thin film of plastic and aluminium, and are sold in huge quantities; this means that they are winding up on the beaches across South East Asia as people cannot easily dispose of them responsibly and there is no incentive for waste pickers to collect them as the material has no value. Countries in South East Asia are often blamed as being disproportionately responsible for the plastic pollution crisis, as more plastic enters into the ocean from this part of the world than any other. However, the culpability of companies that are producing irresponsible products like sachets is rarely discussed.

These companies are locking populations into using throwaway products such as these with no plan for how they will be disposed of, nor are they investing enough in infrastructure to help governments deal with them. By failing to take responsibility for the end of their products' life (something that other industries such as the automobile and electronic waste industries are made to), they are culpable. Plastic pollution, according to the Asia-Pacific Economic Cooperation, is already costing South East Asian countries (ASEAN) more than $1.2 billion in shipping, tourism and fishing as well as impacting the communities living on the coast who have to deal with plastic on their doorstep every day.

"

Tiza Mafira is Director of the Indonesia Plastic Bag Diet Movement (Gerakan Indonesia Diet Kantong Plastik) and works with many groups in Indonesia fighting against the tide of plastic pollution.

Who are you and what do you do?

Tiza Mafira, Director of the Indonesia Plastic Bag Diet Movement.

Why do you care about plastic so much?

It clogs up waterways in my city, Jakarta, and comprises the bulk of waste from our river clean-ups. It is terrifying how it takes hundreds of years to degrade and yet we use it for single-use purposes. Humans have survived for so long without plastic and our dependence on it today seems so pointless.

What can people do to help?

Stop using single-use plastic. Reject and refuse every time you are given a carry bag, a straw, a plastic bottle.

What's the worst example of plastic pollution you've seen?

In and around the Ciliwung River that runs through Jakarta, plastic waste is buried two to three metres deep on the riverbanks, deposited there during high floods and slowly building up layers of plastic and mud. It is impossible to clean up.

What's the best solution to reducing plastic you've come across?

Bans. We learn time and time again from the numerous countries or cities that have banned some forms of single-use plastic, that once you have banned it and are serious about enforcing the ban, the amount of plastic waste plummets. When our organization succeeded in pushing for a plastic bag charge trial that lasted six months in 2016, we saw plastic bag use drop by 55 per cent. Afterwards Banjarmasin became the first city in Indonesia to ban plastic bags, and they saw plastic bag use drop by 80 per cent immediately.

Are there any changes you've made in your life to reduce plastic?

I always carry a reusable bag and a tumbler. I reject plastic bags, plastic bottles, coffee cups, straws, skincare products containing microbeads and food packaged in styrofoam. I co-founded the Indonesia Plastic Bag Diet Movement to campaign against the excessive use of plastic bags and demand the government enact plastic-reduction policies.

What makes you most annoyed when it comes to plastic?

How it is almost impossible to escape because every single product is packaged in plastic. How government officials often shy away from creating plastic-reduction policies because they claim 'the people aren't ready', as if they didn't have the power to change behaviour.

Do you have any top tips for getting rid of plastic?

I work on policy, and I believe that plastic-reduction policies will have a better success rate when implemented gradually, starting from retailer plastics (such as plastic bags and straws, because they are separate items from the product we are consuming and are therefore not essential) to fast-moving consumer goods plastics (such as food or shampoo packaging, because this requires more effort to come up with an innovative redesign of the packaging).

What do you think is the biggest challenge in getting rid of plastic?

Resistance from the plastics and petrochemicals industry.

What do you think is the biggest opportunity in getting rid of plastic?

For Indonesia, the government needs to issue that national charge on plastic bags that they have been planning for so long. For the entire world, make plastic-free bulk stores more popular.

What's the most impressive effort to reduce plastic you've seen – from an individual or a company?

I am extremely impressed by any individual who is living a zero-waste lifestyle. I haven't managed.

If this rundown of the global problem shows anything, it is that it is easy to drown in the flood of plastic-related statistics coming out every day, to lose perspective in the onslaught of bad news stories from around the world. The media is full of new horrendous facts and figures about the crisis, and scientists are racing to produce more research to help us better understand the mess we are in. Politicians and businesses faced with difficult decisions about what to do are struggling to keep pace. Faced with so much information, it can be hard to work out what's what, so here's a quick summary of some of the most important statistics to help persuade your colleagues, friends and family of the need to move away from our throwaway lifestyles.

PLASTIC BY NUMBERS

120 billion plastic bottles made
by Coca-Cola every year

· · ·

38 billion fragments of plastic
found on the uninhabited
Henderson Island in the South Pacific

· · ·

330 million tonnes of plastic
produced every year

· · ·

12.7 million tonnes of plastic
entering the ocean every year

· · ·

500,000 plastic particles per square
metre in a river in Manchester, UK –
thought to be the highest intensity
ever discovered in one place

450 years for a plastic bottle to
break down in the ocean

. . .

111 years since the first plastic was invented

. . .

90 per cent of seabirds have plastic in
their stomachs

. . .

80 per cent of plastic in the ocean originates
on land

. . .

53 years since the plastic bag was created

. . .

1 rubbish truck of plastic enters
the ocean every minute

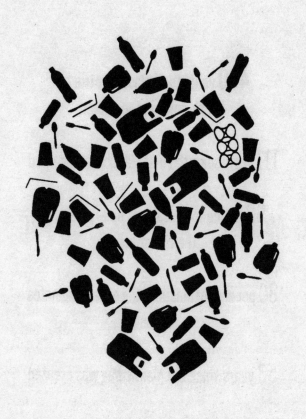

3

STORIES OF HOPE AND SUCCESS – A GLOBAL MOVEMENT AGAINST PLASTIC

Amidst the rising tide of plastic entering our oceans, it's easy to feel a bit overwhelmed. In my job you learn to be resilient. If I didn't, then falling into a spiral of despair would be all too easy – the fact that we have already changed our world almost beyond recognition is a difficult thing to come to terms with. When the pessimism seeps through, though, there are easy ways out. I look at the inspiring people around me – working in all kinds of jobs in all kinds of organizations who care desperately about making our world a better place and work tirelessly to make it so. I look at people at the beginning of their journey, taking their first steps to understanding that everyone can make a difference. And I look to where I can take action in my own life, because nothing beats feeling down at the end of a long day than making plans and actually doing something myself to join the movement of people trying to get rid of plastic.

And it is a growing movement that inspires me every day. Even the most cursory search on the internet reveals stories across the world of passionate people taking action in their own lives, and companies and governments listening to the concerns people have and introducing measures to reduce plastic. Here are some of the most inspirational plastic-related stories I've heard about.

Plastic Bag Bans Around the World

Plastic shopping bags have become a symbol for plastic pollution across the world. With an average lifespan of just fifteen minutes, scientists estimate they might take between 500 and 1,000 years to break down. Like all plastic products, they are a relatively recent phenomenon and anyone over the age of fifty will be able to remember a time before they became so ubiquitous. As a symbol of what is wrong with single-use plastic, and one of the most commonly found items on beaches, it is unsurprising that so many countries and regions around the world have started to ban them. Starting with Bangladesh in 2002, the last few years have seen a wave of plastic bag bans across every continent (except Antarctica, but Antarctica doesn't have a permanent population anyway).

Like all legislative efforts to reduce plastic, these bans require enforcement, but where they are being properly regulated they have been hugely successful. For example in Morocco, the second largest consumer of plastic bags after the United States, nearly 500 tonnes of plastic bags were seized or confiscated after an outright ban came into effect in summer 2016. Country- and city-wide bans on plastic bags are already having an impact on how much plastic enters the ocean.

Other Single-Use Plastic Bans

For some places, banning plastic bags doesn't go far enough and no single-use plastic item is safe. San Francisco, famed for its seafront with colonies of pelicans and sea lions, continues to pave the way with bans on plastic bottles, peanut bags, the foam balls inside beanbags and more. Going even further is the province of Karnataka in southern India where all single-use plastic items have been banned, including bags, banners, cutlery and more. Still in its early stages, the province's government is learning how best to enforce this ban, but the message it sent was powerful: future generations should not have to deal with our plastic waste.

From Antigua and Barbuda banning styrofoam containers, to the EU's recently announced ban on single-use plastics like straws and cutlery – globally, single-use plastics are coming under more and more scrutiny by politicians, and simple bans can be amongst the most effective and immediate ways to deal with the plastic crisis.

193 Countries Admitting There's a Problem

One of the most interesting developments in recent years was the coming together of 193 countries in Nairobi in December 2017 to discuss the issue of plastic. At the end of the summit, these countries issued a common statement declaring the need for urgent action to tackle this growing problem. Although the statement came under some criticism for its failure to deal with specifics and the bureaucratic process it proposed, nevertheless I found it inspiring. One former colleague, well versed in UN negotiations after years working on the climate conferences, once reminded me that a common statement of intent from the UN is nothing short of a miracle. That somehow nearly 200 countries can come together, put aside their differences – which in a few cases extend to all-out wars between some of them – and agree on anything at all is a remarkable example of common humanity. The statement was accompanied by specific pledges from 39 of the countries to take national action to reduce plastic entering the ocean. It is my sincere hope that this statement paves the way to ambitious international action on plastic, but at the very least, it marked the first major international step: admitting the problem.

People Going Plastic Free

From whole families attempting to live entirely without plastic for a week to the 'Plastic Free July' pledge already taken by over 2 million people in more than 150 countries, there are thousands of inspiring stories to be collected about the challenges and opportunities of living without plastic. A plastic-free life may not be possible for most people – time, money, geography and other factors can all play a part in preventing us from fully dedicating ourselves to giving up plastic completely. However, that so many people are giving it a go and documenting their attempts online of what is quite a significant challenge is a constant source of inspiration. For every question you have that isn't answered in the pages of this book, I can guarantee that several of these amazing people will have the answers online across countless blogs, Instagram posts and Facebook pages.

Plastic-Free Blogs

There are plenty of tips in the coming pages about how to reduce the amount of plastic you consume, and how to get those around you to reduce theirs – but if you ever need more information or inspiration, then take a look at some of these fantastic blogs, all giving advice on living without plastic.

Plastic Free July:

A global movement of people trying to live with less plastic in July. Filled with inspiring stories, useful advice and daily motivation – will you take part this July?

Beth Terry – My Plastic-Free Life:

Beth's blog contains an amazing 100 steps to get started on living life without plastic, as well as a super-interesting challenge to you to take pictures of your weekly plastic rubbish as a way of measuring your impact.

Anne Marie – The Zero-Waste Chef:

For those who love cooking but can't find the non-plastic-wrapped ingredients, Anne Marie has plenty of tips and recipes to help go plastic free in the kitchen.

If we continue to share ideas and support each other in our efforts to cut back on plastic, it will make our goal more achievable. I'd love to hear how you are cutting back on plastic and any of the alternatives you have found. Share them with the community at

#BreakFreeFromPlastic

"

Louise Edge is a campaigner at Greenpeace who has spent most of the past few years working on all aspects of the plastic issue. From negotiating in boardrooms to working with impacted communities, she's seen every side of the plastic problem.

Who are you and what do you do?

I'm Louise Edge and I'm an oceans campaigner at Greenpeace.

Why do you care about plastic so much?

Because I see the extent to which plastic is contaminating our natural world. Wherever scientists look today – whether it's in our rivers, Arctic ice or seafood – they are finding tiny plastic particles. And this pollution is being eaten by everything from tiny plankton at the bottom of the ocean food chain to great whales at the top. This is not only killing wildlife, it is also altering their behaviour in some worrying ways. The potential impact on human health from exposure to plastic particles also gives me real pause for thought: this is a material made from petroleum that contains a whole range of chemical additives, some of them toxic – and we are ingesting it.

What can people do to help?

There are lots of small changes we can make in our daily lives and in our communities to reduce the use of single-use plastic packaging, which is one of the major sources of ocean plastic pollution. However, it is big companies and government who have the power to solve this problem, by massively reducing their use of plastic packaging. So we need to let them know we want change – through social media, face to face when we visit stores or meet local politicians and through our purchasing decisions.

What's the worst example of plastic pollution you've seen?

In 2016 I visited Freedom Island, a coastal bird sanctuary in Manila Bay, where every inch of the beach was smothered in plastic packaging. You could dig down for a metre and still find layer upon layer of plastic waste mixed with sand. In the sea the plastic waste bobbed alongside dead fish and birds. It was a profoundly upsetting experience. I joined a group of local volunteers to clean the beach, but it was a thankless task as each day the tide brings in yet more brightly coloured plastic packaging. It really brought home to me the scale of this problem and the fact we really need the big companies who had made this packaging – like Nestlé and Unilever – to act to solve it.

What's the best solution to reducing plastic you've come across?

The best and easiest solution is to just use less: stop over-wrapping products, use reusable packaging wherever possible and when you do need single-use packaging, use a material that can be recycled and doesn't last for ever if it ends up in our oceans or in landfill.

Do you have any top tips for getting rid of plastic?

Plastic is so ubiquitous right now it is hard to completely get rid of it from your life. But there are some easy things you can do. I now carry a reusable water bottle and cup, and reuse shopping bags. I have also switched from liquid soap, shower gel and shampoo to solid versions which aren't plastic packed – Lush have been leading the way on these. For household cleaners I have switched to traditional cleaners like sodium bicarbonate and borax in the kitchen and bathroom. Also, as a fizzy water fan I invested in a 1970s-style home soda maker and saw my recycling bag shrink dramatically!

What makes you most annoyed when it comes to plastic?

When I see images showing the impact plastic is having on marine creatures, like the sperm whale recently washed up dead with a stomach full of plastic, I get very upset. But what annoys me is some of the 'greenwash' you find on packaging, whether it's companies implying that making 'bio'-plastic from plants rather than petroleum somehow makes it 'green' (it doesn't) or putting 'recyclable' logos on to plastics they know full well won't *actually* be recycled in the real world, as the material is too complex and the costs are too high. These types of plastic simply shouldn't be on the market and the greenwash drives me crazy!

What do you think is the biggest challenge in getting rid of plastic?

Producing plastic makes a lot of money for powerful companies like Exxon Mobil and Shell, who along with others are currently ploughing billions into new 'cracking' facilities that produce the raw material for plastic packaging. So we will clearly have some very rich corporations resisting the shift away from plastic packaging and using their budgets to try and persuade the public of their case. I don't think they will succeed as there is such strong public support for change – but it will certainly be a challenge.

What do you think is the biggest opportunity in getting rid of plastic?

It's people saying enough is enough and making companies commit to dramatically reduce single-use plastic packaging and government legislate to ensure this happens. This is happening across the world right now. It's happening because people understand plastic pollution is a problem we have inadvertently created over the past fifty years and it's one we can solve. Our world functioned before plastic packaging existed, so it's obvious that we can exist without it again. It won't be easy as plastic so dominates our lives today, but if we employ our ingenuity it is doable and that's something that really motivates me.

What's the most impressive effort to reduce plastic you've seen – from an individual or a company?

I was blown away by the work of a zero-waste community that I visited in Manila. They formed in response to a waste crisis which saw bin bags literally blocking the streets. They set up a centre for reuse and recycling and a neighbourhood support network which dramatically reduced the amount of waste the neighbourhood sent to landfill to an amazing average of four rubbish bags a day, mostly nappies. All the rest was reused or recycled – amazing!

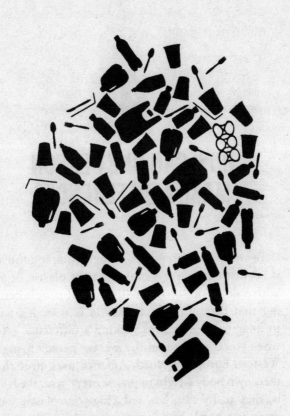

4

HOW CAN ONE PERSON MAKE A DIFFERENCE?

If any of those stories have a single common thread, it is this: that every victory against plastic begins with a single person or small group of people deciding that the time to take action is now. It's hard to imagine your actions making a difference – but when you consider that the average person living in Western Europe or North America uses more than their own body weight in plastic every year, the truth is, they really can. Yes, reducing your plastic footprint by a bottle here, a coffee cup there, may be no more than a drop in the ocean, but the message it sends speaks volumes – and, of course, the ocean is nothing if not countless drops of water.

Every victory against plastic begins with a single person or small group of people deciding that the time to take action is now.

The changes we make in our own life can have far-reaching ripples – especially if we're good at communicating with others about why we're doing it. After all, politicians and company CEOs are just people, like anybody else, and if they hear our stories of why we want to get rid of plastic, and why we think they need to help do something about it, they can be persuaded to act. Humans by nature are social beings – we all have networks extending through our families, friends and colleagues – and with the rise in technology our ability to communicate with our networks has never been greater. As you go through the following chapters and start getting rid of plastics in your own life, remember that one of the most powerful things you can do is talk about what you're doing and why you're doing it. In that way, others may follow!

HOW TO GIVE UP PLASTIC: A PLEDGE

If you're feeling inspired by the work of some of the people and campaigns I've described, then before we move into the next section of the book, getting into the detail about how to give up plastic, think about making this pledge.

Starting today, I pledge to do my best to give up plastic. It's not an easy journey, or a short one, and in many cases it may not be completely possible, but here's me trying to do my best by pledging:

- **To refuse plastic wherever I can, such as not using plastic straws, bags, coffee cups or bottles.**
- **To reduce my plastic footprint whenever possible by choosing non-plastic materials built to last.**
- **To reuse plastic items like containers where I can't refuse or reduce them.**
- **To recycle or repurpose everything else that I can.**
- **To tell everyone I know about what I'm doing to get rid of plastic and encourage them to join me!**

Signed .... Date ....

#BreakFreeFromPlastic

One of the most inspiring examples of people making waves in their effort to reduce plastic pollution are the sisters Amy and Ella Meek, aged fourteen and twelve, who set up Kids Against Plastic and regularly appear in the news talking about the need to clear our lives of plastic.

Why do you care about plastic so much?

As the future generation, plastic pollution is an issue that we are going to inherit so we want to make it as small a legacy as possible. Since founding Kids Against Plastic, we've been working to pick up 100,000 pieces of the Big 4 plastic polluters (single-use cups and lids, straws, bottles and bags) – one for every sea mammal killed by plastic in the oceans annually. We've also encouraged many cafes, businesses, schools and even our local council to become Plastic Clever, meaning they are more discerning users of single-use plastic and promote the use of reusable items. We try to spread the word about plastic as much as possible too, doing talks (such as a TEDx!) and workshops in schools, as well as having our own crew of Kids Against Plastic around the country.

What can people do to help?

Become Plastic Clever! Reduce your use of plastic cups and lids, straws, bottles and bags.

What's the worst example of plastic pollution you've seen?

In a village called Arrochar in Scotland – it's at the end of a tidal loch. A horrendous amount of plastic washes up on their shore. It really brought the issue of plastic pollution home for us, especially seeing the effect of the litter on the locals, who produced none of it. They've almost given up against the unrelenting tide of plastic, but still try to complete monthly beach cleans on the lochside. It showed that the plastic we throw away is impacting all areas of the globe – not just the developing countries we see in photos and on the news.

What's the best solution to reducing plastic you've come across?

Using reusable items – it can even save you money, as most cafes now offer discounts for using reusable mugs!

Are there any changes you've made in your life to reduce plastic?

We avoid single-use plastic, in particular the Big 4 (cups and lids, straws, bottles and bags).

What makes you most annoyed when it comes to plastic?

Plastic packaging on items that don't even need wrapping! We especially get annoyed when fresh fruit and vegetables are covered in layers of unnecessary plastic, and us customers don't even get the option to purchase it loose!

Do you have any top tips for getting rid of plastic?

Start small! Refusing single-use plastic can actually have a large positive impact – more so than you might imagine.

What do you think is the biggest challenge in getting rid of plastic?

The fact that we're all too reliant on this convenience packaging. It's going to take work to break our plastic habits!

Giving Up Plastic at Home

Armed with all the reasons why we need to get rid of plastic, now it's time to get into the nitty-gritty. The next few chapters are a guide to lots of ways you can reduce your plastic footprint. But first, a top tip – don't try to do it all at once! Just like making twenty New Year's resolutions, you're setting yourself up for failure. So, take it slow and steady, introducing new changes and making new decisions about what products to use (or not use) week by week. One of the best places to start is the bathroom . . .

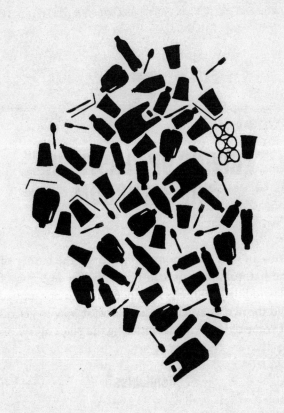

5

GIVING UP PLASTIC IN THE BATHROOM

Look in your bathroom cabinet or at the corner of the bathtub and you can probably see an array of plastic-filled containers – all destined to be emptied and thrown away. Let's take a look at how to get rid of some of the plastics cluttering the place up.

Refillables

Whether your bottle of shampoo, conditioner or hand cream lasts a week or a month, it probably comes in a plastic dispenser of some kind, used once and then replaced. As the desire to get rid of plastic grows, there is increasing demand for refillables from people like you and me who would rather either buy in bulk and refill when empty from a large container stored away or take our smaller containers back to a shop and have them refilled directly.

For the reusable dispensers, most homeware shops will stock bottles with pumps that you can use; alternatively, you could look at reusing any nice old glass bottles such as the blue bottles that Neal's Yard sells its products in. You may be lucky enough to live near a shop that will refill them for you. Companies like Ecover have refill stations in many independent shops and many markets also have stalls for refilling your own bottles. As a lot of us don't have access to these facilities just yet, instead you could find a wholesaler near you or online to buy from in bulk. If you have the space, consider getting a 5 or even 10 litre bottle of your chosen lotion, which will last many months (or longer), and use that to top up your refillable pump.

Some shops like the Body Shop used to offer a refill service but dropped it due to lack of customer demand, so if you do want to refill at your favourite shop then it's always worth letting the manager know that they risk losing your custom if they don't introduce a refill service. Unfortunately, however, despite the fact that refill stations are relatively easy for shops to introduce, they remain quite unusual for the time being. If your search for a way of refilling your dispensers proves fruitless, then you could consider looking for non-liquid soaps and scrubs that don't require plastic packaging.

Bars

Increasingly, people trying to get rid of plastic in their bathroom are turning to solid soaps and shampoos. From high street shops like Lush to countless online vendors, soap and shampoo bars are becoming mainstream again. Make sure they don't come wrapped in plastic, but come in a reusable tin or box instead, and the switch away from liquids might be the easiest way for you to empty your bathroom of plastic. Deodorant bars and creams are also becoming more widely available and a quick search for natural deodorants on Etsy should show you a range of options.

Sponges

Having a plastic-free scrub couldn't be easier. Plenty of doctors actually recommend against using washcloths or sponges in the bathroom because unless cared for properly they can cultivate bacteria. However, if you can't imagine a shower without a sponge or scrub, then buy a loofah (a kind of dried fruit) – available pretty much everywhere.

Avoid Microbeads

Of course, what your toiletries come wrapped in might not be the only problem, but also what they contain. Until recently few people had heard of microbeads, but as countries across the world got wise to this pernicious addition to our creams and cosmetics word spread and soon governments everywhere were banning them. Used for a variety of purposes such as exfoliating, the use of microbeads became quite widespread in a short space of time as companies moved away from more natural alternatives like ground apricot kernels. They can be found in plenty of everyday products including toothpaste, sun cream, make-up, face and hand wash and more.

In countries that haven't already taken the plunge or where the ban has yet to take effect, here are some things to look out for to make sure you're not unwittingly washing thousands of these polluting beads down the drain every time you wash your face or clean your teeth. Beat the Microbead is a coalition of organizations (www.beatthemicrobead.org) trying to end the use of all microbeads and they have previously produced a useful list of products and companies that have pledged not to use microbeads on their website.

However, if you're just browsing or want to see if what you already have in your cupboard might be a part of the problem, then have a look for these ingredients on the back of the packet:

- Polyethylene (PE)
- Polypropylene (PP)
- Polyethylene terephthalate (PET)
- Polymethyl methacrylate (PMMA)
- Polytetrafluoroethylene (PTFE)
- Nylon

If you do find that a product you've already bought contains microbeads, then my best suggestion is to post it back to the vendor and ask for a refund. They might not give you your money back, but at least they'll get the message that customers don't want to be accidentally polluting the ocean at their bathroom sink.

Cotton Buds

Shortly after the microbeads ban, several countries including Scotland and France moved straight on to banning any cotton buds with plastic in them, soon to be joined by the UK. When Waitrose banned them across its UK branches it estimated it would save 21 tonnes of plastic – not bad for such a small product!

If you can't imagine not using these little buds for cleaning out your ears or removing make-up, then when you're searching for a non-polluting cotton bud choose ones with bamboo or paper stalks instead of plastic. Johnson & Johnson, one of the largest manufacturers, has pledged to get rid of plastic in their cotton buds. Unfortunately they've only made the pledge in some countries, so it's worth ringing them to ask when they plan on introducing it where you live if they haven't already, as there's no excuse for such half-hearted measures.

Make-up

Make-up packaging is a little problematic. There are a few brands, such as Fat and the Moon, that sell items like blusher and foundation in tins, but at the moment it can be hard to avoid plastic in your make-up bag. If changing make-up brand really doesn't work for you then it's time to kick up a fuss. Write to your favourite brands and let them know their packaging isn't living up to their customers' expectations – that you need them to do better by innovating to find new forms of packaging.

Make-up removal, however, is much easier. There's no need to buy single-use make-up removal pads which often come wrapped in plastic, and are also sometimes made using plastic fibres; instead try the variety of plastic-free alternatives on the market such as the reusable cotton pads made by Sin Plastico or the compostable Konjac sponges made from vegetable root.

Lip Balm

Whether it comes in a tin or a compostable cardboard tube, plastic-free lip balm is fortunately quite easy to find. Many mainstream brands offer their products in tins as well as tubes, or have a look online for one of the hundreds of options to soothe your lips guilt-free.

Brushing Your Teeth

Dental care is a little trickier. Not only might you have to keep an eye out for microbeads, to find toothpaste not in a tube takes a little more effort, and a toothbrush without plastic bristles even harder. If you do want plastic-free oral health, there are a couple of options. For toothpaste there are two brands easy to order online that package their toothpaste in glass jars: Truthpaste and Georganics (don't worry, they still keep your teeth clean!). Alternatively, you could go for some vintage dental powder – used for centuries and just as effective at cleaning your teeth. There are plenty of brands out there selling it in glass jars or you could even make your own using this recipe by Kathryn, who runs www.goingzerowaste.com.

Make Your Own Tooth Powder

- ¼ cup of xylitol: it's a natural sweetener. It prevents bacteria from sticking to the teeth and neutralizes the pH to help avoid tooth decay.
- ¼ cup of baking soda: a very mild abrasive (less abrasive than commercial toothpastes) that dislodges plaque on teeth, breaks down stain-causing molecules and neutralizes pH.
- ¼ cup of bentonite clay: draws out toxins, contains calcium and is often used to help remineralize teeth.

Stir together. Avoid using metal with the clay, it will deactivate. I use a wooden spoon and store it in a glass mason jar. This will leave you with clean breath and tastes like nothing. The sweetness in the xylitol cancels out the saltiness of the baking soda and the clay is very neutral.

Bamboo toothbrushes are becoming more common and this option (or another kind of wooden-handled toothbrush from a sustainable source) at least transforms most of the toothbrush into a biodegradable alternative. Many of these toothbrushes make claims of plastic-free bristles, but my advice is to always read the small print. Truly compostable bristle alternatives are nigh on impossible to find and bamboo or other wooden-handled brushes are still using some form of plastic bristles (of these, Brush with Bamboo is the most advanced in reducing the percentage of plastic in the bristles). For the really ambitious (and non-vegan), Cebra Ethical Skincare, a German brand, makes toothbrushes the old way using boar bristles – though of course this may not be to everyone's taste.

For those that use a toothpick after meals, there are plenty of reusable options made from titanium. For flossing I would recommend Le Negri, which is entirely plastic free, or Dental Lace, which offers refillables by post (although the floss line itself is made of a form of plastic). You could also use silk thread as recommended by some plastic-free bloggers.

Hair Removal

In case your decision to get rid of plastic doesn't coincide with a decision to grow out your body and facial hair, then shaving is another part of your bathroom routine where single-use plastics can be found in abundance. First up is your razor. Disposable razors with multiple blades are no good for the environment; single-use razors are even worse. Buy yourself a safety razor and a pack of blades. It might be scary at first, but honestly – the effect is the same. Even better, it doesn't take long to make the money back with each blade lasting a couple of weeks, and the body of the razor will last as long as you take care of it. Most chemists and high street shops still stock razors like this.

For men's shaving soap, it is easy to find alternatives online packaged in a traditional wooden bowl. For cream, the easiest way to go plastic free is to forgo shaving-specific cream and use a bar of soap instead to lather up. However, have a look online such as at the Living Without Plastic website (www.pfree.co.uk) and there are a couple of shops and brands that produce creams or shaving-specific soap bars in reusable (or no) packaging.

Shaving is certainly the easier route to plastic-free hair removal, although it is possible to remove plastic even if you prefer to wax. Ordinary waxing relies on plastic strips and entirely synthetic materials. MOOM are a company dedicated to plastic-free waxing (or sugaring) and have produced strips and organic wax all sent to you in non-plastic packaging. Alternatively, many plastic-free bloggers have opted for the more DIY approach and you can search online for the hundreds of recipes to make your own sugar-based wax, which will simply wash off any cotton strips you use it with.

Tampons

On the subject of plastic-free periods, Natalie Fee, founder of UK-based organization City to Sea, is the expert. Natalie points out that 'people are totally shocked to discover their pads and tampons have plastic in them'. Given that the average woman will use between 12,000 to 16,000 tampons in her lifetime, it's not surprising that tampons and their applicators are a fairly regularly found item washed up on beaches – particularly after storms when sewers may have overflowed. In City to Sea's fantastic short film on the subject called *Plastic Free Periods?* they revealed that the average disposable menstrual pad contains as much plastic as four carrier bags. It may sound obvious, but it's worth reiterating that tampons, even compostable ones, should not be flushed away – they weren't designed to go down the toilet!

One easy like-for-like option is to switch to the compostable products made by Natracare and available globally, whose tampons aim to deal with unethical and unsustainable practices. Or Tampon Tribe, who will even donate a day pack of supplies to homeless women for every month's worth that you buy. Alternatively, you could opt for a reusable moon-cup whose central claim is that 'you only need one'. It is more expensive up front, but as it's reusable you'll have made your money back in just a few months.

Down the Loo

It's not just tampons that are flushed away only to reappear on our beaches and in the ocean. Wet wipes are also an all too common sight on the seashore.

Wet wipes, which come in plastic wrapping, are often made using plastic fibres. If you're a regular user of wet wipes, have a look at the make-up removal section to see whether you might be able to replace them in your daily routine or indeed get rid of them altogether and use a cloth instead. However, if you do feel the need to use them then the first rule of using wet wipes is throw them away in the bin, not down the toilet!

Toilet paper may be made of paper, but what it's wrapped in frequently isn't. Who Gives a Crap, Pure Planet, EcoLeaf and Seventh Generation are all companies that deliver paper-wrapped toilet paper to your doorstep. What's more, they deliver in bulk so you're less likely to get caught short!

If you bought it in any high street shop, chances are the brush you use to clean the bowl is either entirely plastic or at least has plastic bristles. You can order a plastic-free toilet brush from Plastic Free Life (the source of so many plastic-free everyday items) that uses pig bristles – or, for vegans, Boobalou makes one using plant-based bristles instead.

Now that you've been through your bathroom, how about writing down the plastic-free plan you've found to meet your taste, budget and location in the chart on the following pages – take a picture of it and share online for others to follow your example!

ITEM	PLASTIC-FREE PLAN
Shampoo	
Soap	
Hand wash	
Shaving cream	
Razor	
Deodorant	
Sponge	
Lipstick	
Foundation	
Blusher	
Other make-up products	

Toothbrush	
Toothpaste	
Lip balm	
Make-up removal	
Tampons	
Toilet paper	
Toilet brush	
Other	

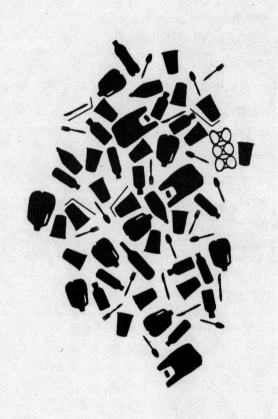

6

GIVING UP PLASTIC IN THE BEDROOM

Microfibres and Clothing

It comes as a surprise to most people that the clothes they wear are one of the greatest sources of plastic in the ocean. Minuscule strands of clothing, normally made of nylon or polyester and much finer than a human hair, are shed from our clothes every time we wear them, wash them and, of course, when we throw them away. The rise of fast fashion means that as a cheap and easy-to-use material, polyester makes up around 60 per cent of the clothing material we wear, with an estimated 61 million tonnes of synthetic fibres being manufactured in 2016 according to UN figures.

A report released in 2017 by the International

Union for Conservation of Nature[2] estimated that between 15 per cent and 31 per cent of all plastic pollution comes from microplastics. The authors estimated that an average person living in Europe is responsible for dumping the equivalent of 54 plastic shopping bags in the ocean each year. The figure for North America rises to 150 per person per year. The report goes on to explain that, globally, over a third of this plastic enters the ocean as a result of us washing our clothes. At less than a millimetre long, the microfibres are so small they slip through our washing machine drainage systems. A fleece jacket could be responsible for releasing as many as 250,000 microfibres according to a study undertaken at the University of California, Santa Barbara.[3] I find it galling that as someone who enjoys spending time outdoors cycling and kayaking in the elements because I love nature, the equipment I use and the clothing I wear can be some of the most polluting.

It is not ridiculous to ask how, if these fibres are so small, they are actually harming the ocean. The answer – as is so often the case with this fast-growing

2. https://portals.iucn.org/library/sites/library/files/documents/2017-002.pdf

3. Hartline, N. L., Bruce, N. J., Karba, S. N., Ruff, E. O., Sonar, S. U. and Holden, P. A. (2016), 'Microfiber Masses Recovered from Conventional Machine Washing of New or Aged Garments', *Environmental Science & Technology*, Vol. 50, No. 21, pp. 11532–8.

problem – is that we have yet to understand the true impact.

A fleece jacket could be responsible for releasing as many as 250,000 microfibres.

What we do know is that although invisible to our naked eye, these synthetic fibres may still look like tasty treats to zooplankton such as krill, tiny shrimp-like crustaceans. Animals like these form the base of the food chain in the ocean, being eaten in vast quantities by bigger zooplankton, fish and marine mammals like whales. In this way microfibres can pass up the food chain, accumulating in huge quantities the further up you get, and eventually could even end up on our dinner plate. In addition to this, in the same way as larger plastics can block up the stomachs of birds and whales, microfibres can prevent other zooplankton like copepods from being able to digest the algae on which they depend.[4]

The International Union for Conservation of Nature estimated that between 15 per cent and 31 per cent of all plastic pollution comes from microplastics.

What to do then? If these synthetic fabrics are so widespread and the pollution they cause largely unseen, a solution can be hard to imagine. Here are a few things that you can do to help stop such large quantities of microfibres entering the ocean.

4. https://www.researchgate.net/publication/236926420_Microplastic_Ingestion_by_Zooplankton

Shopping

Buying fewer clothes

I know the feeling, the weather changes and you want the outfit to match or a zip breaks on your jeans and it's easier to buy a cheap new pair. Clothes can be found so cheaply now that we often forget it is a relatively new phenomenon to be able to head out to the high street and buy them in large quantities on a whim, without necessarily considering the impact on the environment or the people who have made them. Making an effort to reduce the amount of clothes you buy, either by repairing old ones or making do with last year's colour scheme, is a simple, effective way to reduce the amount of microfibres in the environment (and save money at the same time!). The longer the clothes you already own stay in use, the kinder you are being to the environment.

Buying fewer new clothes

New is not always nice for the environment. Next time you're out shopping, try checking out charity shops and vintage stores for second-hand gems; after all, fashion comes in cycles. Also, although they are still synthetic and so would require the washing guidelines below, you could check out some of the clothes now on the market made using recycled plastic, such as Pharrell Williams' line or any one of hundreds of small start-ups doing everything from high fashion to gym wear.

Buying fewer synthetic clothes

Before you purchase, have a look at the label to see what it's made of. Where possible, see if you can opt for more natural materials like wool, cotton and silk. Unfortunately it's true that these materials will often be more expensive, but the flipside is that well-made clothes from natural materials should last a long time. If you're buying outdoors clothing, check for suppliers like Fjällräven and Patagonia that are trying to reduce the amount of microfibres in their products. Also, if you can bear it, avoid buying fluffy clothing and materials like fleeces as these can be some of the worst offenders in the washing machine.

Make your voice heard

In Chapter 11 we'll look at how to campaign in more detail, but if you're in a shop and trying to do the right thing by searching for clothes made of natural materials and everything you like contains synthetic fibres, then do what any customer displeased with the service they are receiving should consider doing: complain! The more people who raise their voice – either in private with the shop manager, via email to the customer service team or in public on social media – the more these companies will listen to the fact that people don't want to be responsible for polluting our oceans through the clothes they buy.

Washing

Do you really need to wash that?

Only wash synthetic clothing when you have to. I'm often guilty at the end of the day of just putting my clothes in the laundry basket, when often after a day in the office they could easily be worn again.

Wash your clothes smarter

Research by clothing company Patagonia suggests that when you do wash your synthetics, the following measures can reduce the amount of microfibres they shed:

- Washing at lower temperatures (ideally a cold wash)
- Making sure you have a full load
- Using a lower spin speed and a shorter cycle
- Using fabric softener and liquid laundry detergent

Buy a washing machine with a microfibre filter

Although not yet on the market, hopefully the growing demand will soon see the introduction of washing machines with in-built microfibre filters, such as those being developed by the EU-funded project called Mermaids. If you're purchasing a new washing machine in a few years' time, see if you can get one with such a filter.

Use washing detergent with less plastic packaging

Although not strictly related to microfibres, your washing detergent can also be a source of plastic waste. Instead of individually wrapped plastic capsules, use washing powder in a cardboard box; if you prefer liquid detergent, then buy it in large quantities to reduce the number of plastic bottles you're using.

Bedding, Carpets, Furniture and Mattresses

Of course, clothing isn't the only fabric in the bedroom. Soft furnishings may also use synthetic fibres – however, as they are washed far less often, they represent a much smaller piece of the puzzle. When it comes to bedding, the same rules apply as with clothes: make sure to choose sheets and covers made using natural materials like cotton and silk.

If you're really keen to put the ethical icing on the cake, you could go a step further and source the rest of your soft furnishings from companies using recycled plastic. Companies like Weaver Green have a huge and increasingly affordable range of cushions, carpets, bags and even dog beds all made using recycled plastic bottles (it's amazing how much plastic can feel like wool!). It isn't hard to go even further than this and look at beds, mattresses and duvets made from recycled materials by companies such as Nimbus and

Silentnight. I've no doubt in the years to come such products will continue to become more widespread and accessible as the awareness grows of the need to recapture plastic before it enters the environment.

7

GIVING UP PLASTIC IN THE KITCHEN

As the end destination for our most frequent purchases, the kitchen can be quite a daunting place to get rid of plastic, particularly when faced with such limited choice from the most common supermarkets. Accounting for nearly half of our individual plastic consumption is packaging of the products we buy. If you speak to any of the big supermarkets about the need for them to reduce their plastic packaging, a common statistic you'll hear in response is that although plastic is a problem, without plastic there would be 40 per cent more food waste. Yet a recent report by Friends of the Earth Europe found that plastic packaging has increased in direct correlation with food waste: European household food waste nearly doubled between 2004 and 2014, whilst plastic packaging rose by more than 25 per cent.

The report also found that many packaging practices actually result in increasing wastefulness as brands and retailers try to grab customer attention with over-packaging and two-for-one or buy-one-get-one-half-price deals that lead to people buying more than they need. Consumer research shows that most people would prefer to buy a single reduced item rather than the several they are encouraged to, and yet whether it's to promote a brand or for other savings, many retailers continue to use too much packaging in this way. It's clear from the study that the route to ending food waste, much like the route to ending plastic pollution, requires many solutions. For example, if supermarkets didn't feel the need to have constantly full shelves regardless of customer demand for a product or insist on selling only the most perfect-looking vegetables, would waste figures be quite so high?

Supermarkets are businesses fighting each other in a dog-eat-dog world – retail is fiercely competitive and monthly sales figures are pored over by teams of experts desperate to find clues as to how to get one up on their closest rivals. To help with this one-upmanship the supermarkets employ whole teams focused on innovation, coming up with ideas to make products more appealing and accessible to consumers. They say that necessity is the mother of invention – well then, let's make it necessary for these supermarkets to innovate and start working much faster to find alternatives to individually wrapping every product they can. In the UK, the supermarket chain Iceland is leading by example by pledging to go plastic free by 2023. By raising our voice and complaining, even if it's just a note in the comments box of your online shop, we can help make sure these teams of highly skilled engineers and designers are given the order to start thinking outside the (plastic) box and come up with new ways of delivering products to consumers that don't rely on packaging that we have no effective solution to dispose of in such quantities.

European household food waste nearly doubled between 2004 and 2014, whilst plastic packaging rose by more than 25 per cent.

In the meantime, here are some of the many ways you can reduce the plastic in your kitchen.

Going Shopping

Preparation

Before you head out of the door, go through a quick mental checklist of what you're heading out for. One of the most common ways I end up using lots of plastic is making last-minute purchases on the go or by rushing out quickly without making a list of what I'm getting or from where. A small amount of preparation and thinking through your shopping trip can often be a great way to reduce how much plastic you use whilst you're out.

Ditch the Plastic Bags

Your final check before stepping outside the house on your way to the shop has to be for your reusable bags. These could be supermarket-bought 'bags for life' or cotton tote bags or even just a rucksack. Over 500 billion plastic bags are used every year globally – that's at least a million a minute – and yet replacing your plastic bags with a reusable alternative could be the easiest thing you do to get rid of plastic. If you're doing your shopping online, then make sure to write in the comments box of your order that you do not want the items delivered in plastic bags. It's possible the supermarket will ignore you, but the more people writing this, the more likely it is they'll start including it as an option.

Replacing your plastic bags with a reusable alternative could be the easiest thing you do to get rid of plastic.

Where to Shop

Make It Work Wherever You Shop

One option could be to carry on at your favourite or most convenient shop – to avoid their most over-packaged goods and to put pressure on them as a regular customer to do more to reduce their plastic footprint. For example, you could ask to speak to the manager or the customer services representative in the shop (or on the phone), or if you're on social media then take a photo of the packaging you're most frustrated by and tag the shop in your picture.

Using Social Media

One easy way to vent your frustration whilst making companies think twice about using so much plastic is to take pictures of the worst offenders and, making sure to tag the social media account of the relevant shop or brand, put your picture on Instagram, Snapchat, Twitter, Facebook and any other social media platforms you're a member of, using the hashtag:

#BreakFreeFromPlastic

Everyone I speak to can think of a time they've been absolutely outraged at the amount of plastic something they've bought is wrapped in. Whether it's individually wrapped pieces of fruit, shrink-wrapped again on top of a plastic tray, specially crafted plastic holders for bite-sized chunks of chocolate or tiny quantities of pre-chopped vegetables in an oversized plastic bag, it is not hard to find examples of packaging gone too far. Early in 2018, amid the plethora of plastic-related commitments from retailers and brands across the world, UK-based supermarket Marks & Spencer was found to be stocking 'cauliflower steak' – two slices of cauliflower wrapped in plastic – for £2, twice the price of an entire unwrapped cauliflower. A justifiably irritated shopper tweeted a picture of it, and after becoming the subject of much ridicule the supermarket dropped the 'steak' altogether. The moral: if you see something that has no place on our shelves, that defies the laws of common sense, then share it with your friends, family and followers online, and you might help some of these companies see how stupid their packaging can become when left unchallenged.

Many mainstream supermarkets also stock loose fruit and vegetables, sometimes even 'wonky' produce that doesn't meet their usual exacting aesthetic standards, and by choosing this you can help contribute to reducing food waste as well as plastic waste.

Shop Local and Independent

Alternatively, you could start looking at other shopping destinations that don't use so much plastic. Local grocery stores, health food shops and markets are often some of the best places to get loose fruit and vegetables, which you can either put in a paper bag or just straight into your own reusable one. Local butchers, delicatessens, bakers and fishmongers are also more likely to wrap your fresh food up in paper or to put it in any reusable container you give to them. Food being sold fresh is far less likely to come in plastic packaging because the point of buying fresh is to eat fresh and so there is no need for any kind of packaging to store it. If you don't have a convenient independent shop, then many of the larger supermarkets also have fresh meat and dairy counters – see whether they will wrap your cheese or meat in paper or even in your own container.

Local, of course, may not always be an option – perhaps in your area it's too costly or the choice is too limited. In this case you could explore home delivery options. Whether it's vegetable or meat box schemes such as Farmbox or Riverford, or your supermarket of choice has agreed not to use plastic for your order, home delivery for many people can be a good way to reduce waste.

The Future of Plastic-Free Shopping

In an inspiring development, shops dedicated to helping customers go plastic free are springing up across the world. From Zero in Fremantle, Australia, to Earth.Food.Love in Totnes, UK, independent retailers as annoyed as the customers with the over-packaging of their goods are trying to offer an alternative vision of a future without plastic.

What to Buy

Fresh Produce

So that leaves the question, what should you be buying to reduce your plastic footprint? As already explained, fresh and local can often be the best way to reduce your plastic packaging. Options available to many include the market in your town or village, or the local delicatessen where you can build a relationship with the owner and talk to them about ways to reduce plastic packaging, as well as ask them to wrap your goods in whatever you choose. If there's one thing that's become even clearer to me in researching this book, it's that going completely plastic free in our shopping is not an option available to everyone. It is so dependent on where you live and how much time you might have to do research, so although many of the options presented here represent the ideal, even simple things like picking the loose vegetables in the supermarket over the plastic-wrapped multipacks is a fantastic way to reduce your plastic footprint.

Dried Goods

Dried goods can also be a part of your shopping where you can easily reduce your plastic footprint. This is because dried goods are the simplest of your kitchen cupboard to buy in bulk. Think about buying in 5 or

10 kilogram quantities. If your local shop or supermarket doesn't sell in bulk, then try finding an online wholesaler, such as Infinity Foods or Naturally Good Food, who cares about reducing their packaging.

To make this work, you'll need some good storage containers in the kitchen. Either use old glass jars that you've washed out or seek out some nicer ones in a charity or homeware shop. You can then buy large bags of dried pasta, pulses, grains and beans and transfer smaller quantities into the jars in your kitchen cupboard. Get some clips or rubber bands as well to tie up the larger bags to store at the back of your cupboards or in the shed. Properly stored in a dry environment, dried goods like pasta can last for three years and most kinds of rice even longer; just empty these foods into the smaller containers as and when you need to top up rather than buying them in smaller plastic bags every time.

Dried goods can be a part of your shopping where you can easily reduce your plastic footprint.

If you want to reduce all kinds of waste, not just plastic, then think about bulk-buying foods that you normally buy in tins as well, which can often come with plastic wrapping around them or use plastic labelling. If you buy these products dried then you'll have to get used to soaking them for a few hours before you want to cook, but the upside is you'll probably save quite a bit of money as buying dried pulses and beans in bulk is almost always the cheaper option. In addition to this, although aluminium tins are more regularly recycled than most plastic packaging, the production process has its own set of problems that it's good to avoid where possible. A growing number of shops around the world are beginning to sell dried goods in smaller quantities in paper bags (or even in whatever container you supply), bringing us the best of both worlds.

Some of the worst cases of excessive waste come in the form of multipacks – plastic packages inside plastic packages, with the contents often going to the bin as well, contributing to the growing food waste problem. When going shopping, be sure to try and buy only what you need rather than getting tempted by offers of more for the price of less that are likely to be no more than repackaged versions of the same thing you can get further down the shelf. Of course, if you do need a larger quantity or to save money on a particular product, look out for a single large package (sometimes even in paper or cardboard rather than plastic), instead of several smaller ones.

Make Your Own

If you're a keen cook or have the time, then one of the most effective ways to give up plastic is to buy loose, fresh ingredients and then make your own. Snacks in particular are one of the areas of our shopping more likely to create plastic waste: crisps, chocolate, even pre-cut fruit and vegetables for dips. By making your own guacamole or your own energy bars you can save money and use your own reusable non-plastic containers. Homemade snacks are incredibly easy to make in bulk and store – search online for your favourite recipes.

Avoid Unrecyclable Plastic

Once you're doing your shop there are a few kinds of packaging, such as styrofoam, polystyrene and PVC, that should be put on your red list, because it is so unlikely that they will get recycled; instead they'll end up polluting our environment or in landfill. It's ridiculous that more supermarkets haven't done the right thing and completely eliminated this kind of unrecyclable packaging.

Ready-meals, fresh fruit and meat all frequently come wrapped on top of a black plastic tray. Although the kind of plastic they are made of is often actually recyclable, the sorting machines at recycling plants cannot distinguish these trays against the black conveyor belts and so they are sent to be incinerated or to landfill instead. Research shows that using an alternative pigment could cost less than a penny per tray extra, so if you're annoyed that your favourite meal is contributing to the billions of these trays that aren't being recycled every year, then make sure to let the company producing it know.

If you have no choice, as will often be the case, but to buy products wrapped in plastic, then keep an eye out for the internationally recognizable recycling symbol, three arrows in a triangle, and try to choose items with this on. At least this means there is a good chance you will be able to recycle it somewhere. Packaging without this symbol should be avoided. City to city and country to country, what can get recycled varies so much that to attempt a list here would be impossible. However, it's easy to do a bit of research. Go to the website of whichever author-ity organizes your local waste collection or that runs your local dump – somewhere on the site there will be a list of the different products you can recycle with them, ideally with accompanying logos.

Drinks

The drinks we buy account for some of the most commonly seen items on our beaches and in the ocean. In addition to the now well-known problem of plastic bottles and bottle caps, which are dealt with in the next chapter, six-pack plastic rings, coffee pods and even teabags are all playing a major part in polluting our environment.

Coffee pods

The rise in coffee machines in kitchens has unfortunately led to a growing market in single-use coffee pods, such as those marketed by Nespresso and other companies. These pods frequently use plastic for the main shell, with an aluminium top – separating these two materials is not something the vast majority of recycling centres are willing or able to do.

If you haven't yet bought a coffee machine geared towards single servings then it's worth reconsidering as the relative environmental impact compared to regular French press, instant, filter or any one of the many, many other forms of making coffee is likely to be much bigger. In Hamburg, for example, all state-owned buildings have banned the use of coffee pods in an effort to reduce their waste; in the US a campaign against one of the biggest manufacturers, Keurig, who have only committed to having 100 per cent recyclable pods by 2020, has seen their sales drop substantially.

However, if you're addicted to the convenient

morning fix then there are a couple of options to reduce your impact. In first place are compostable pods, meaning that after use you can stick them in with your food waste. A number of companies like Lavazza and Dualit have developed compostable pods, and more companies are set to follow. Have a look online or ask your local supermarket to help find compostable coffee pods that are compatible with your machine. If you don't have any luck then the next best alternative is to look for a way to recycle the pods you are using, by choosing pods made of easily recycled materials (they should have the recycling logo on them). In some cases you can return yours in person or by post to the shop you bought them in; Nespresso even arrange home collections.

Teabags

The same compostable rule applies to teabags. Unfortunately in the general move to use plastic in everything, many tea companies started using plastic to seal the bags, leading to many of us polluting the environment unwittingly as we threw them in with our food waste. After more than 200,000 gardeners signed a petition to Unilever, the tea brand owned by the company, PG Tips, made the commitment to move away from using plastic. If there is no 'plastic free' label on the brand that you choose, search online to see if they've made any statement, and if they haven't then you can ring, email or tweet their customer services team to double-check. If you're really keen to minimize waste, then consider making your morning cuppa with loose leaf tea – use a teapot with an in-built infuser or a tea leaf ball to keep the leaves out of your mug.

Milk

Although no longer as widely spread, in many places around the world milk is still available in glass bottles, which are returned after use, to be washed and used again. Do some research to find out whether such a scheme is available in your area. If you live rurally, are there any farms nearby that will refill your bottles for you?

Cooking and Cleaning

The apparatus we use in the kitchen can also occasionally be harming the planet. Where possible try to use second-hand glass containers like old jars or long-lasting metal ones. This will also help you reduce your plastic use on the go – as you'll see in the next chapter, takeaway food is one of the biggest culprits in over-packaging. Get containers with lids so that you don't need to use cellophane to store food or have a look at companies like Bee's Wrap or Eco Snack Wrap.

Another easy thing to buy in bulk is your cleaning fluids. Washing-up liquid and detergent are both relatively easy to find in large quantities or – as with your toiletries – there are some shops that will allow you to refill your own container, be that an old bottle that you're reusing or a prettier second-hand one that you've repurposed. Ecover is one of the best companies out there, stocked in many mainstream shops, online and in partnership with many markets and smaller outlets offering refillable options. What's more, their packaging is made mostly out of recycled materials, attempting to close the loop on what packaging they do use.

For scrubbing your kitchen it's also easy to swap to the same loofahs you've switched to in your bathroom – these natural sponges are compostable rather than just going in the rubbish and ending up in landfill. Alternatively, using dish and floor cloths that you wash regularly is much better than single-use wipes or towels that won't necessarily be recycled. Try to buy cloths made using natural materials like wool and cotton as a way to reduce the microfibres that will get released when you wash them.

When it comes to the kitchen, there is no greater authority than the Zero-Waste Chef, Anne Marie. If these pages aren't exhaustive enough for you and you want to go the extra mile and fully expunge all plastic from your food preparation, then you can use her book or blog that includes recipes specifically featuring ingredients that can be sourced from non-plastic packaging. If you're not much of a cooking fan and tend to eat out or get takeaways, then speak to your favourite takeaway about whether they'd consider switching away from plastic packaging or maybe even use the containers you provide them.

Now that you've been through your kitchen, how about writing down the plastic-free plan you've found to meet your taste, budget and location in this chart – take a picture of it and share online for others to follow your example!

ITEM	PLASTIC-FREE PLAN
Plastic bags	
Fresh fruit and veg	
Meat and fish	
Dairy	
Dried goods	
Snacks	
Coffee	

Tea	
Food storage	
Washing-up liquid	
Sponges and cloths	
Takeaways	
Other items you purchase regularly . . .	

Who are you?

My name is Bonnie Wright, I'm an oceans activist,
director and actor.

Why do you care about plastic so much?

I guess for me it's a mixture of things. I surf, so
I'm in the water quite often, so I see plastic pollution
firsthand. Both here in LA and when I'm visiting
countries where the infrastructure is too poor to
dispose of waste, it's distressing to see more and
more people relying on the convenience of plastic.
When I'm looking at water and all I see is plastic, I
just know it's going to be there for ever. That concept
of it not breaking down really shocks and saddens me.
But it's something that we can actually do something
about. I think there are many issues in this world that
sadly, due to the politics around them and the
injustices wrapped up in them, you can't really do
much about, but, with this one, my actions can help
create a shift in what we're buying.

Yes absolutely! When I go round the UK and give talks, people are impressed by the idea that they might actually be able to do something about this environmental problem. You said it's heartbreaking knowing it will not break down, but what's the worst example of plastic pollution that you've come across?

When I was aboard Greenpeace's ship the *Arctic Sunrise* we did a lot of plastic trawling off the Atlantic coast. That was interesting because you have bottles floating in the water, or something quite large, then you see all of the tiny pieces of plastic that have broken down into microplastics. Doing that trawl right in the middle of the Gulf Stream when we were going from the Bahamas to Miami, it looks seemingly beautiful and blue, but then we dropped the trawling net and after just one hour on one section of a giant ocean, we already had loads of plastic inside. That really hit home because it was the tiniest percentage of the ocean, but the crew said that there wasn't one time that they trawled and didn't bring up plastic. That moment put it into sharper perspective than seeing rubbish on a beach.

What makes you most annoyed when it comes to plastic?

I think that plastic, through unnecessary packaging, has made people stop trusting the natural protection that fruit and vegetables have. For example, the skin of a fruit is much better protection for that piece of food than all the packaging that it is wrapped in, so at what point did we think that wrapping it in all of this packaging was good for us? I think that the psychology behind waste and cleanliness and what we deem *clean* has been a bad thing in terms of our reliance on and our love affair with plastic.

What do you think the biggest challenge in fighting plastic pollution is?

We can all choose to use our recyclable cups or our reusable bags, but at the end of the day, even though some of us do have the luxury of choice, a lot of people don't. So I think it goes back to the need to pressure corporations to put this at the top of their agenda. It's really a question of who is going to be bold enough to use the money and resources they have towards reimagining materials and packaging. It is hard, because I don't think people are going to stop consuming food and drinks that are packaged, so how do we work within this consumer society and make it more circular? I think it's a challenge as the convenience is a reality for people who have bigger issues to deal with; I mean, they might have only just enough money to buy their lunch. Despite not wanting plastic packaging, they aren't really able to choose something else. I think the question is: how do we make big companies eradicate plastic and use a better material?

Do you see an opportunity there?

Yes, I think the beauty of living in a society where consumers are such a focus is that we have the power. So, if it becomes cool and fashionable to package things only in reusable materials, then companies will start to do that because they don't want to lose customers to, say, another fast food company that might already be changing their ways.

What about the most impressive effort that you've seen from an individual – is there any one person or group of ordinary people that comes to mind?

I've always been so impressed by friends around me who start changing just one thing. It happened with me when I thought, 'I'll start using a reusable water bottle' and 'I'm never going to buy a plastic water bottle.' It just takes a moment's thought to trigger that domino effect when you realize that it's everywhere. A lot of friends will send me pictures saying, 'Oh look, I got this reusable cup!' I love friends sending me their proud moments.

Are there any changes you have made? I know we've spoken about reusable cups and bags but people are always hungry for more tips.

I agree. I'm slowly starting to clear all plastic out of my house, but it's hard and it takes time. My favourite things are reusable – such as fabric fruit and vegetable bags. I also have bulk food bags, so I get things – like my granola that I have for breakfast or flour for baking – in one of these reusable bags so I don't have to buy them in the packaging. I always have my reusable cup for coffee, tea or water. Toiletries are a big thing, and I found this company that has a subscription service whereby they send you shampoo and conditioner, and you send back the metal containers.

Is there any other message that you want to put out there in terms of what people can do to help, or ways to stay inspired?

If I had a message, it would be that yes, it is overwhelming and it is a really big issue, but these small changes that you are making are significant. It can be hard, so just choose one part of your household – such as food, or cleaning products, or toiletries – to tackle first. A friend of mine, Lauren Singer, who has a blog called *Trash Is for Tossers*, lives a completely zero-waste lifestyle, and in a year the amount of rubbish she creates fits inside just one mason jar. She also has a great company called the Package Free Shop, and they are an amazing resource to help you reuse more and waste less. I would advise people to see the power in their local restaurant or coffee shop; don't be afraid to talk to people about their use of plastic, or maybe ask them to give an incentive to sell more reusable cups through a discount. Have gentle, open conversations with your local businesses to find out if they are thinking about these things, because in the past year or two it has become something that everyone is aware of, and we can already see changes happening.

8

GIVING UP PLASTIC ON THE GO

Walking down the street, the litter we are used to seeing is the remnants of our lives spent on the go. Crisp packets, food cartons, coffee stirrers and bottles scattered across the pavement serve as a reminder of the speed with which we now live our lives with no time to stop and enjoy the moment. Throughout this chapter, try and think of the ways that you might be able to reduce the amount you buy on the go.

If you head out of the door in a hurry and forget your plastic-free essentials but are desperate for a bite to eat or a caffeine fix, then don't beat yourself up about it. Even if you forget for a whole week, that's still fifty-one weeks of the year when you're not contributing to the problem of plastics on the go, which is no small achievement. Lifestyles built around our ability to pop into a shop and pick up a meal or snack at a moment's notice are not going to change overnight. If we are serious about giving up plastic we have to gradually unlearn these habits – they are, after all, recently acquired. Luckily this can be a phased project as there are plenty of quick but very effective ways to reduce your plastic consumption on the go.

Plastic Bottles

In the not too distant past, plastic bottles were a rare commodity compared to reusable glass bottles. Around 500 billion plastic bottles are sold every year, and growing – that's 20,000 a second. If placed end to end, they would reach halfway to the sun. Although the rise of plastic bottles brought with it some benefits, such as reduced carbon emissions transporting them due to their lighter weight, it was an ill-thought-through replacement given the lack of a plan for what would happen after they had been used once. In places still using glass bottles, which are still reasonably widespread in some parts of the world such as many African and Latin American countries, especially for milk and juice, if the switch to plastic has not yet begun then producers and retailers should probably hold back.

It's worth remembering that on-the-go plastic has made life significantly easier for many people who, for example, cannot drink easily without a straw or for whom safe water from the tap is not a given. Our own efforts to give up plastic on the go, and to persuade our friends and family to do likewise, are to be much praised, but before pointing the finger at a stranger or another country's practices, it's worth considering the other reasons they may be unable to give up plastic just yet. Here is Jamie Szymkowiak, co-founder of disability rights organization One in Five, writing about why getting rid of plastics shouldn't suck for disabled people:

In responding to public concerns about plastic pollution, a number of transport providers, cinema and restaurant chains and sports venues have, understandably, committed to phasing out the provision of plastic straws. A few companies have replaced plastic straws with paper or metal alternatives, whereas some are withdrawing all straws from public display until a suitable alternative is sourced or withdrawing straws altogether. As politicians continue to exercise their clout in the anti-plastics debate, it's important that we consider the wider implications of a ban, particularly for disabled people, as we move towards eradicating single-use plastic straws.

The average plastic straw is cheap, flexible, can be used for drinking cold and hot beverages and is readily available. For some disabled people these attributes are vital for independent living. It's important to note that the umbrella of 'disability' includes people with different needs and impairments, and that it's the universal accessibility of the plastic straw that makes so many disabled people anxious about an outright ban.

Disabled people can take longer to drink, therefore a soggy paper straw increases the risk of choking. Most paper and silicone alternatives are not flexible, and this is an important feature for people with mobility related impairments. Metal, glass and bamboo straws present obvious dangers for people who have difficulty controlling their bite, as well as those with neurological conditions such as Parkinson's. Some disabled people use straws when drinking coffee or eating soup, yet most of the alternatives, including the leading biodegradable straw, are not suitable for drinks over 40°C. In addition, reusable straws in public places are not always hygienic or easy to clean – would you drink through a straw that's been passed around the public?

One of the most common rebuttals from non-disabled people is that disabled people should just bring their own straw. Think about that for a moment. In addition to our Blue Badge, medicines, bank card and phone, we must also remember to carry a straw at all times just in case we get thirsty?

Then there is the cost. According to Scope, disabled people in the UK already face extra costs of £570 a month related to their impairment or condition. Passing yet another cost on to disabled people isn't suitable if you accept that society bears a responsibility to make the world more accessible for everyone. After all, environmental justice without social justice isn't justice at all.

What can we do?

I'm part of a disability rights group, One in Five, that is calling on manufacturers to produce an environmentally friendly flexible non-plastic straw that is suitable for hot and cold drinks – and we need support from non-disabled people too. When companies are discussing their needs with suppliers, they're unlikely to buy four or five different straws; therefore we need a universal solution.

During an episode of BBC One's *The One Show*, the managing director of Iceland Foods, Richard Walker, exhibited a clear, paper-based and recyclable alternative to the plastic film that covers many of their frozen meals. Although it's still in the developmental stages, this demonstrates that companies will respond to consumer demands and that an environmentally friendly straw that meets the needs of disabled people and doesn't pollute our oceans is not beyond our capabilities.

I think it's important to point out that not one disabled person I've discussed this topic with is against the principle of banning unnecessary single-use plastics. In fact, many of the disability rights activists I know also champion animal rights and the need to protect our environment for future generations.

As we move to ridding our oceans, beaches and parks of unnecessary single-use plastics, disabled people shouldn't be used as a scapegoat by large corporations or governments, unwilling to push suppliers and manufacturers to produce a better solution. Instead, we must all work together to demand an environmentally friendly solution that meets all our needs, including those of disabled people.

Reusable Bottles

Reducing our dependence on plastic bottles is crucial to ending our throwaway culture. Getting yourself a good reusable water bottle to carry around may be one of the most important steps to giving up plastic in your life. In Chapter 11 you'll find the tools to help you campaign for more water fountains in your area, and to make sure local cafes and restaurants help reduce plastic by offering to refill your bottle. However, if you regularly buy a bottle of water a day then just by refilling it from the tap before you leave the house each morning you'll already be reducing your plastic footprint by 365 bottles a year. It's true that some places can be snooty about refilling your bottle – if you're met with a frown when you ask for some water from the tap then make sure to remind them how easy it would be to accuse them online of polluting the ocean, with a picture of the plastic bottle they're making you buy.

In the US over 1,500 plastic bottles are used every second!

If you're concerned that a steel flask doesn't suit the image you're going for, then don't worry. As the movement to give up plastic has grown, so too has the variety of reusable bottles available. From the practical Klean Kanteen to the trendier S'well or Chilly's, or the huge selection you'll see in any camping shop, finding a water bottle to suit your look has never been easier.

Soda Maker

However, given that just in the US over 1,500 plastic bottles are used every second, we still need to find more ways to refuse and reduce our use of plastic bottles. One way may be to purchase a carbonator. There are many brands of soda maker out there, helpfully collated into one place on the website www.sodamakerclub.com. They will fizz up your drink and you can flavour it how you like, using syrup or even more natural flavourings.

When Throwaway is the Only Option

On those occasions when neither soda maker nor refillable bottle is meeting your needs and buying a throwaway drink is the only option (we've all been there), then in order of priority try to make your choice according to the following:

1. Pick a drink in a container made of more easily recyclable materials like a cardboard carton, a can or a glass bottle.
These don't represent fully sustainable alternatives as they're still part of the same system that sees us invest a large amount of energy into something that we use once and then throw away, but they at least provide more of a guarantee that they will be recycled.

2. Pick a brand that uses more recycled plastic to make their bottles.

By using recycled plastic they are reducing the demand for new plastic to be produced and incentivizing the market for used plastic, meaning that it's more likely not to go to waste. Companies like Naked juices or Resource water are both leading the way with 100 per cent recycled content bottles. Don't be put off by the cloudy-looking plastic of these bottles – having completely clear plastic is one of the excuses companies use for not going 100 per cent despite the fact that most of us really don't mind what colour the bottle we buy our drinks in is (although a company called Ioniqa recently partnered with Unilever to try and revolutionize plastic recycling by being able to recycle it to 'good as new').

3. Pick a brand using 100 per cent recyclable plastic.

This is really at the bottom of the pile simply because there is absolutely no reason why every single brand isn't making bottles that are 100 per cent recyclable. If you can't recycle the materials, you shouldn't be putting them on the market. Most major companies have committed to going 100 per cent recyclable within the next ten years – which, given how many plastic bottles they're making every year, isn't too impressive.

Finally, if you do get a plastic bottle, make sure to dispose of it responsibly. Rather than leaving it perched on top of an overflowing bin on the pavement, stick it in your bag and put it in the bin at home later (hopefully the recycling bin!). If you're lucky enough to live somewhere with a deposit return scheme, whereby you pay a small deposit for the bottle when you buy it, which you can get back when you return it to a shop later, then be sure to use it. These schemes are hugely effective in reducing the number of bottles that enter the environment.

Coffee Cups

It is a ubiquitous morning sight – commuters clutching their coffee cup desperately drinking in the caffeine before the slog ahead. Most of us, myself included, didn't even realize there was a problem with these cups until relatively recently. I normally refused to take a lid, but thought the cup was okay as it felt like cardboard. Then in summer 2016, celebrity chef and environmental campaigner Hugh Fearnley-Whittingstall released a series called *War on Waste*. As well as uncovering much about the global food waste scandal, he and his team also lifted the lid on how our addiction to convenient coffee was creating mountains of plastic waste. Despite their cardboard exterior, coffee cups contain a thin film of plastic on the inside, rendering most of them unrecyclable. Of the 2.5 billion coffee cups that British people use each year, only 0.25 per cent are recycled; Starbucks alone uses over 4 billion coffee cups a year with only nominal efforts in some countries to reduce this staggering number.

Of the 2.5 billion coffee cups that British people use each year, only 0.25 per cent are recycled.

The simplest, most effective and only way that I would endorse to give up plastic coffee cups is to get yourself a reusable cup (or a couple if you're like me and prone to losing things and want to leave one at work). Coming in all sizes, price ranges and colours, reusable cups are no longer the preserve of well-off hippies – practically every service station and even plenty of coffee shops sell them at reasonable prices. The most famous brand, KeepCup, has sold millions in over thirty countries. What's more, many of the leading coffee shop chains offer a discount if you bring your own cup so you could make the money back. You can even get foldable ones now that pack into a tiny bag. Refuse the plastic stirrers and ask for a metal teaspoon instead, and soon these unnecessary products should be made a thing of the past.

Cutlery

Walk along a beach in any tourist destination and you're likely to find plastic forks and spoons half buried in the sand, ready to be washed away to sea, where they will take hundreds of years to break down. Plastic cutlery, often sealed in its own plastic bag, has become a normal part of life on the move. To add to the wastefulness, if you're anything like me you get given a knife, fork and spoon but rarely use anything other than the fork to scoop out the salad beneath the plastic film. Carrying around your own cutlery (and let's face it, you probably don't need to carry around all three) and refusing to take the plastic stuff when it's offered to you is a great way to reduce your plastic footprint. Either borrow a set from your kitchen drawer or buy a smaller more portable set like those found in camping shops. If you're using chopsticks then just keep the last plastic set you got given with your meal, and you're all set to go. If space really is an issue in your bag, how about getting a Spork – knife, spoon and fork in one.

Plastic Bags

There is not much more to be said about plastic bags beyond the fact that they should be consigned to the history book as a once useful, now redundant fad.

From clogging drainage systems to being eaten by turtles that mistake them for jellyfish, they are the symbol of plastic pollution everywhere. Estimating the volume of plastic bags in circulation is a near impossible task, but it's not unreasonable to assume they may be the most commonly found single-use plastic item worldwide. One by one countries are realizing the need to end their use and it must only be a short while before the rest of the world follows suit. Wherever you live you probably already have a selection of reusable bags perfectly suited to going to the shop, or if you use a car for your weekly shop then grab some cardboard boxes from the supermarket to help load your shopping rather than using bags.

Straws

Ever since a YouTube video emerged of a straw being slowly, painfully extracted from a turtle's nose, plastic straws have been in the spotlight. Aside from the exceptions explained by Jamie at the start of this chapter (page 117), plastic straws have no place in modern society. Wetherspoon, one of the largest chains of pubs in the UK, recently announced their decision to end the use of plastic straws in their premises, opting instead for compostable alternatives only. However, as with almost all the on-the-go items listed so far, the best option is to just say no. When you're ordering your drink at the bar or in a restaurant, try to remember to say that you don't want a straw. If drinking through a thin tube really is your preference, then there are plenty of reusable dishwasher-proof alternatives to be found online.

There is not much more to be said about plastic bags beyond the fact that they should be consigned to the history book as a once useful, now redundant fad.

Convenience Food

From sandwiches in plastic cartons to salads in plastic boxes, yoghurt pots and fruit salads beneath a plastic film – rushing to the shop during our break to get a quick lunch is often the time of day when we can end up having to throw away the most plastic. When you're hungry and in a rush it's hard to hunt for the plastic-free alternative and so one of the best ways to give up the plastic that comes with convenience food is by preparing to avoid it altogether. Just as you buy in bulk, think about cooking in bulk too, storing food in the freezer and using it for packed lunches or quick dinners later in the week. It'll save you money and reduce your dependency on convenience food and drink that is far more likely to have a degree of waste attached.

I feel your pain at being reminded of the obvious – in a week where I'm working long hours the last thing I need to think about in the evening, on a precious weekend off or on a rushed lunch break is how to reduce plastic by preparing for the days better. Once you get into the habit, though, it will soon become second nature.

How about writing down the plastic-free plan you've found to meet your taste, budget and location in the chart on the next page – take a picture of it and share online for others to follow your example!

ITEM	PLASTIC-FREE PLAN
Plastic bottles	
Coffee cups	
Cutlery	
Plastic bags	
Straws	
Convenience food	
Other	

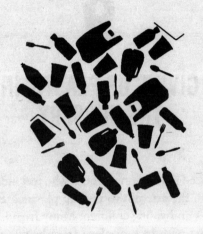

9

GIVING UP PLASTIC IN THE NURSERY

I've noticed that amongst those who feel most belea-
guered in the attempt to give up plastic are parents
of babies and young children. Many friends already
struggling with lack of sleep from being woken up
through the night by their new baby have expressed
to me how frustrated they are by the amount of plastic waste that seems to be part and parcel of having
a child. If ever there was an excuse for choosing
convenience it is the screaming child you're trying to
pacify, so if you do resort every now and then to the
throwaway option then don't ruminate on it. By
using reusable or non-polluting options the rest of
the time you're still doing a huge amount to reduce
your plastic footprint.

Nappies

In the US alone, an estimated 27.4 billion nappies are used every year, with more than 90 per cent of them ending up in landfill where they will take over half a millennium to break down. As well as plastic, nappies also use large quantities of wood pulp and require a lot of energy to produce. The alternative: reusable cloth nappies like those from Bambino Mio or Wonderoos. Well known to our grandparents, these have only improved in time and are no longer anything like as cumbersome as they once were. For the best explanation of how to care for reusable nappies I can recommend Life Without Plastic's blog post on the topic.[5] Though no one can deny the convenience of not having to wash endless rounds of dirty diapers, if you ended up using even half the number of disposables, that would still be a 50 per cent improvement on using them all the time and so well worth doing. For those occasions when disposables are really saving your sanity, keep an eye out for the compostable ones on the market.

5. https://lifewithoutplastic.com/store/blog/plastic-free-reusable-organic-cotton-cloth-diapers-healthy-baby-planet/

In the US alone, an estimated 27.4 billion nappies are used every year, with more than 90 per cent of them ending up in landfill where they will take over half a millennium to break down.

Dummies/Pacifiers

Dummies are a regular item found at any landfill site. If your child uses one, then have a look at Hevea's range of dummies and teethers. Made using natural rubber, the company also makes bottles and bath toys.

Glitter and Decorations

Beautiful sparkling glitter – the favourite decoration of so many children (and so many adult festival goers) – is made of thousands of tiny pieces of plastic ready to be washed down the drain or blow away in the wind. At this point I hear the chorus of voices rising up to call me the fun police; in my quest to give up plastic I've gone too far in demanding an end to glitter. Well, if you're a glitter addict, and I don't blame you, then I have good news: there are more sustainable alternatives. They're not perfect as although they are biodegradable they may still contain non-compostable ingredients that mean they cannot be broken down into organic matter; however, Lush, Eco Glitter Fun and Glitter Revolution are three companies leading the charge to plastic-free sparkles.

Toys

Whether it's peer pressure at school, the outrageous marketing techniques of toy companies or just an innocent visit to the toy shop, chances are your child has their sights set on a toy that you and I both know will last a few weeks before getting broken, out of date or just plain boring. Keeping pace with the toys in vogue is a tricky business.

Buy to Last

There are beautiful handmade toys still being made, which are built to last for generations. If you have the means, then thinking about fewer, higher quality purchases like these is worth it for your own child, but also those that may come after. Bella Luna Toys and Loubilou are two companies that ship beautiful, sustainable toys worldwide.

Buy Second-Hand

Toys go out of date quite quickly as children grow up and grow tired of the same electric car sets; however, what's old to one child may be brand new and exciting to yours, so have a look at eBay or in charity shops for second-hand options that you can then give away or sell once your household tires of them.

Buy Recycled Plastic or Plastic-Free Alternatives

A childhood stalwart, Lego have started a new range of plastic bricks, opting to switch for sugar cane instead. They have set the target of 2030 by which time they will only use 'bio-plastic'. Unfortunately bio-plastic, although made from plant-based material, is still plastic and if it enters the environment will behave much the same way as ordinary, fairly indestructible Lego bricks behave. From the perspective of stopping plastic entering the environment, the fact they encourage customers to pass on unwanted bricks rather than throw them away is more important and indeed the number of adult Lego fans I know who still own much of their childhood set says a lot for what an investment Lego can be. I hope that in the near future they might consider a return scheme, helping to play their part in closing the loop on what is still a plastic item. For slightly older children, I really like Bureo's skateboard made using recycled ocean plastics.

Children's Parties

One of the most common complaints when it comes to giving up plastic with children is children's birthday parties. The mounting pressure in the run-up to the big day to have goodie bags, treats and a room filled to the brim with disposable decorations can be unbearable. Try not to stress out or bow to the overwhelming feeling of guilt at not living up to the expectations of other children and parents.

Make Your Own Plastic-Free Alternatives

If you have the time, explore a website like Pinterest, which is packed with tips on how to hold a plastic-free kids' party (or, indeed, any kind of plastic-free party – weddings are a strong feature of Pinterest). Another website called Instructables has endless instructions on how to create beautiful non-plastic decorations. Here are five ideas for plastic-free party accessories:

- Bunting made from old material
- Pom-poms made using a ball of wool and a cardboard disc
- Get a reusable fabric 'Happy Birthday' banner (lots of options on Etsy)
- Make cookies or cupcakes as a take-home party treat
- Ask for washing-up volunteers so you don't need to use throwaway cups, plates and cutlery

Hold a Plastic Amnesty

If you get on well with the parents of your child's friends, then consider holding a plastic amnesty where you all agree ahead of the school year that this year parties will be more sustainable. You could even get together to buy one lot of decorations and share them – after all, most children simply want a party exactly like the last one they went to.

Wrap Less, Wrap Sustainably

So much wrapping paper that we use is coated in a plastic film that means it can't be recycled. Make sure you buy non-laminated wrapping paper – it may not be glossy but there are plenty of nice designs out there. Alternatively, given that in a recent poll 50 per cent of people said they would prefer an unwrapped gift over using plastic-coated wrapping paper, give yourself a break and don't bother wrapping it at all.

The list of ways to reduce plastic in your child's life is endless and I'd recommend exploring blogs like Life Without Plastic, My Plastic-Free Life or Zero Waste Living for tips.

From skateboards made of ocean plastics to reviving your own childhood train set from the attic – every household is different and I'm sure you can find the solutions to help your family give up plastic.

Now that you've been through your nursery, how about writing down the plastic-free plan you've found to meet your taste, budget and location in the chart on the next page – take a picture of it and share online for others to follow your example!

ITEM	PLASTIC-FREE PLAN
Nappies	
Glitter	
Toys	
Party decorations and goodie bags	
Wrapping paper	
Other	

10

GIVING UP PLASTIC IN THE WORKPLACE

One of the areas of our lives where we have the most power is our workplace. Whether that's formal power in a senior position or informal power because you see your colleagues every day and they can't escape your passion for getting rid of plastic, campaigning in the workplace can be a really effective tool in helping others see the value in giving up plastic. This chapter looks at three ways your workplace can help give up plastic.

Change Behaviour

By this point in the book if you're starting to make even a fraction of the changes suggested, then your colleagues may already have noticed your efforts to give up plastic and have started to ask questions. It may be that simply explaining the issues to a couple of your colleagues, giving them a few of the facts, maybe even sending them an article (or a copy of this book), is enough to persuade them to join you in giving up plastic. More likely a bit more explicit campaigning will be required. Every workplace, even virtual ones, is likely to have some kind of informal space – a noticeboard, canteen or lounge – where people go in their breaks or to catch up with colleagues. Try making some signs to put up, or writing a post for the staff mailing list, about ways to reduce plastic at work. Start with encouraging people to give up the top five: plastic bags, bottles, coffee cups, straws and cutlery. Use a statistic from this book to grab people's attention about the scale of the problem and point them towards an alternative.

Crucial to persuading your colleagues to alter their behaviour is getting the tone right. No one likes to feel nagged or attacked, so when you make signs or speak with them try to use language that feels inclusive and inspiring. For example, instead of putting up signs saying 'Takeaway Cups Not Welcome Here', which is bound to get some people's backs up and may lead to a minor backlash, you could use language like 'Let's Create a Plastic-Free Office' and underneath use icons of the products you most want your colleagues to give up. Instead of using instructional or patronizing language when you invite colleagues to a talk, such as 'We should all be doing better' or 'Say no to straws', try to be more welcoming than challenging. You can use questions like 'Have you ever wanted to use less plastic?' or 'Want to find out how you could give up plastic?' or use humour to make people smile. This way they will get drawn in and then you can let them know about all the different ways they can be a part of the movement. Keeping a focus on solutions helps people stay positive and feel like they are able to do something.

Offer Plastic-Free Products

If you work somewhere big, then think about whether any businesses might send you some free reusable products or at least give you a discount if you bought in bulk for all your colleagues. When Sky made the decision to have a plastic-free workplace by 2020 as part of their Ocean Rescue campaign they launched the initiative by giving every employee a reusable water bottle – could your employer do the same? Speak to your office manager. Chances are they'll be on your side as there's nothing office managers like less than mess, and plastic cluttering up the place is probably a thorn in their side. Could they join you in encouraging employees to give up these easily replaceable plastic products?

Organize a Lunchtime Talk

Another way to gain interest could be to organize a lunchtime talk by a plastics campaigner or expert. Search online for a local group campaigning to reduce plastics like a Greenpeace group or Friends of the Earth – I'm sure their members would be delighted to come and give a talk about why we need to give up plastic. Before I started the plastics campaign at Greenpeace I invited a friend who had worked a lot on the circular economy and reducing waste to come and give a talk about plastic pollution and the turn-out was brilliant – people from across the building came to hear more about the problem and what they could do to help.

Plastic-Free Competition

If you work somewhere with lots of teams, think about ways you could turn giving up plastic into a bit of friendly competition. One day a week or one month per year you could run a competition to see which team can produce the least plastic waste, weighing the bin bags at the end of the challenge. If you start to get lots of interest from colleagues, then organize a bring-and-share plastic-free lunch. Not only will it be a good chance to get to know each other better, but you can brainstorm all the ways that your workplace can give up plastic.

Your Work's Procurement Policies

Take a walk around your office and look out for all the throwaway plastic being used. If you have a canteen, do they use plastic cutlery? Are there only plastic cups at the water cooler? Wherever you see it, make a note of what the item is. Perhaps you could even do the walk around with your office manager or with a couple of colleagues. Ask questions about why they have chosen to use plastic in that instance – is it convenience or is it simply lack of imagination? Speak to those who make the decisions and ask them to look at alternatives. If they are resistant, do some research or use this book to find non-plastic options that you can present to them.

If they really won't budge or if you just can't get their attention, then think about escalating your campaign. Set up a petition at work and ask your colleagues to sign – you could discuss how to get more signatories at your plastic-free lunch. If you can show the decision makers not only how to resolve the problem but also the fact that people in the building want it, then there's a good chance they'll listen and respond to what you're asking. It might feel strange or uncomfortable to speak up at work, so think about asking for support from a union rep. Many unions have green representatives who really want to help campaigns like yours, as well as having plenty of experience in campaigning for better working conditions.

If your building is having any refurbishment or renovation done, speak to those in charge about what environmental considerations they are taking into account. Have they looked at using products like Mohawk's new Airo carpets, which are entirely recyclable (not many people realize that most industrial carpeting is made of unrecyclable plastic, creating huge amounts of waste every time a building is refitted). Another brilliant initiative came out of a partnership between the largest carpet tile company in the world, Interface, and the leading conservation organization, the Zoological Society of London. They came together to form Net-Works, an initiative working in communities and with local micro-finance organizations in the Philippines and Cameroon to collect discarded fishing nets that the communities have no means to recycle and transform them into carpet tiles. In a recent refurbishment project in the Greenpeace office, our facilities manager decided to install sound absorption panels, a common feature in many offices, which are made entirely out of recycled plastic by a German company called EchoJazz. The possibilities for those with the time or mandate to research and invest in amazing new sustainable materials in building projects is endless and an entire book could focus exclusively on this.

Shout it from the Rooftops

Finally, if your workplace does start to move in the right direction, think about ways they could become a plastic-free champion. Sky's announcement to go plastic free, later followed by the BBC, was a huge step forward in increasing the attention of office managers around the country as to how they could also reduce the plastic footprint of the buildings they oversee. If your workplace is prepared to go to these kinds of lengths, then be sure to help them shout out loud and clear about why they're doing it to help encourage other companies to follow suit. They could put it on their social media accounts or make a sign that your customers will see. If you work for a bigger company, then speak to the communications team about making the most of the good news. None of us likes to be left behind and the same is true of businesses – where one goes, others will follow.

11

GIVING UP PLASTIC IN YOUR COMMUNITY

Long gone are the days when the decision makers in society could rest on their laurels, isolated from the everyday experience of those whose lives they had so much control over. Modern technology means anyone can be a part of a global movement; anyone can speak truth to power, and make sure they listen. This chapter equips you with the tools to be a part of that wider change in society. Our ability to organize locally, nationally and internationally has never been stronger – and it will take this kind of combined effort to make the necessary people sit up and listen. Throughout this chapter, as I go through the various ways you can make a change in the community, remember that the most valuable asset you have is your own experience. You can use this book or search online for the answers to technical questions or for the right statistic to back up your arguments, but the most effective method for persuading anyone is to tell them a story from the heart. Nothing beats the authentic voice of someone who has experienced the bad side of the problem they are trying to solve.

Where to Start

If you don't know where to start and the whole thing feels completely bemusing or intimidating, then the best thing to do is seek out like-minded people. Find a local group campaigning on plastics and get in touch – they might have a regular meeting or have organized an event in your area. They might be a group of concerned local people or part of a national organization like Greenpeace. Have a look on the noticeboard of your local cafe, in the pages of your local paper or online. Given how prevalent the issue of plastic pollution is, there is likely to be someone in your area doing something about it. If you're lucky enough to have a choice of groups, look at what it is they are aiming to achieve and try to pick the group most focused on reducing plastic overall, such as through campaigning for bans on products, rather than groups exclusively focused on picking up litter.

That said, litter can be a fantastic way to involve other people in the mission to give up plastic as it's something that everyone can relate to. It can be an effective tool to help everyone experience first-hand the damaging impact of plastics. If you are able and have the energy, think about attending or even organizing a local clean-up. Though we understand that the real route to getting rid of plastic lies in reducing our use of it overall, it's important to make sure we do something to care for the places we love as well. Clean-ups can be a brilliant way not only to help the area you live, work or holiday in look the way you want it to, they can also be a great tool in helping other people in the neighbourhood find out about plastic and start getting active to do something about it.

Remember that the most valuable asset you have is your own experience.

The Marine Conservation Society (MCS) coordinates beach clean-ups around the coast of the UK, encouraging local people to come together to show some love for their surroundings. As part of the charity's Beachwatch programme, volunteers collect data at the same time as clearing rubbish, helping to build a picture of what kind of litter washes up most often on our shores. To help you organize a safe and enjoyable clean-up, they've put together this step-by-step guide, which you could easily adapt for a local park if you don't have a beach to go to. If you've got any further questions, to find out what's already happening in your area or to get more information, be sure to check out their website at www.mcsuk.org/beachwatch.

BEACH CLEAN GUIDE

Planning Your Clean-up and Litter Survey

There are just a few steps involved in organizing a clean-up and survey.

Before the Day

1. Find a beach or park and check online to see if there are already clean-ups organized there.

2. If you're heading to a beach, be sure to check the tide times – best to plan an event for about four hours after high tide, and not on an incoming tide. Use the tides to pick a date and time for your event. Tides for Fishing is a good website for this or ask the local authority for specifics.

3. Contact the beach or landowner (often the local council or if not they can usually help) to get permission to hold your beach clean and survey on their land.

4. While you're on the phone, find out who's responsible for collecting rubbish from the beach and talk to them about where to leave all the extra rubbish you'll have collected. You can also ask if they have any equipment you can borrow.

5. Do a risk assessment for your beach. The MCS website has lots of tips on how to do this and you could ask the local authority or beach owner if there is anything in particular to watch out for. Visit the beach again shortly before the event to make sure it's all up to date.

6. Now you can get advertising! Print off posters. You can even do press releases and send them to local newspapers (turn to page 184 for advice on writing a press release). Resources to help you can be found on the MCS website.

7. Think about a sign-up page, either a Google form or in the UK you can use the Beachwatch website. It's a good idea to email sign-ups about a week in advance and remind people what to bring (suitable clothing and shoes, water, food, sun cream, thick/gardening gloves) and where to meet you.

You're ready for the big day!

On the Day

So you've followed all the steps and you're ready for your clean-up and litter survey, but what to do and bring on the day? Don't worry, this checklist will have you covered.

Things to bring:

- ❏ Your risk assessment for the beach.
- ❏ Pens and paper – for everyone to record all the litter they collect.
- ❏ Bin bags!
- ❏ Any beach-cleaning equipment you have access to (you might be able to borrow some from the council or landowner). A good pair of gardening gloves is sufficient, but you could also bring litter pickers and bag hoops (these hold bin bags open on a windy day).
- ❏ Clipboards to lean on when recording the litter.
- ❏ Weighing scales to weigh your haul at the end of the day.
- ❏ If you can, bring a first-aid kit, a sharps box (for syringes and needles) and a bucket for glass and non-medical sharps.
- ❏ Survey forms (if you're in the UK use those provided on the MCS website; if you live elsewhere then use theirs as a template).
- ❏ Parental consent forms for under 16s.

Get to the beach in plenty of time so you can do a recce for hazards and set up your 100-metre survey section and so you're there for your keenest volunteers. When everyone's arrived, start the briefing.

How to Do a Briefing

It's really important to brief your volunteers before they get cleaning. They need to know what they're doing, why and how – as well as all the risks and how to keep safe. Here are some of the key points you should cover:

- **Introduce** yourself.
- Provide a bit of **background**: talk about the issue of marine litter and why recording the litter you find is so important. You can mention anything you know about your local area too – what you've found on your beach in the past, any local issues or stats, whether there are any litter campaigns in your area.
- Go over the key **health and safety issues**, including any risks particular to the beach.
- Explain the **survey form and how to use it**.
- You could run a competition for the highest number of items recorded – NB this is number of items, not weight. This encourages people to follow the correct methodology by picking up as much as they can find, and also to tally up their survey forms, saving you some time at the end!
- Ask the volunteers if they consent to you documenting the beach clean through photographs online.
- Give them a **time to meet** back at the start point.

During the Clean

- Be available to help people identify the litter they find and locate it on the survey form.
- If you've got sharps boxes, buckets for glass etc. or a first-aid kit, keep these with you and look out for anyone in need.
- Take photos, capturing what you find and your team of beach cleaners who have provided photographic consent. Share your experience – post on social media and share your day with the online community.

After the Clean

- Weigh and count the bin bags, and count how many volunteers took part.
- Ask if anyone found anything unusual.
- Find out what the majority of items were made of (likely to be plastic).
- Thank them for their help.

Before you leave the beach, make sure the litter is left in the place you agreed with the cleaners and fill in the front page of your survey summary form: number of bin bags, number of volunteers, weather conditions, etc. Share your survey form with MCS or your local organization.

Back at Home

Have a cup of tea and a metaphorical pat on the back from us – you've done an important and great thing today, and it's really appreciated. Pretty soon, after a few events, you'll get some regulars coming along who know the drill and can even help new volunteers who join in. In time, some people might adopt their own beach to get cleaning. It's a ripple effect and you really can make a difference for your whole coastline. Make sure to share pictures and stories about your day online, and encourage all those who attended to do the same when you get in touch later to thank them.

#BreakFreeFromPlastic

"

I've never met anyone as enthusiastic about cleaning up beaches as Catherine, who works for the Marine Conservation Society in Scotland. Here are some of her thoughts on plastic pollution.

Who are you?

Catherine Gemmell, Scotland Conservation Officer for the Marine Conservation Society.

Why do you care about plastic so much?

Through my role at the Marine Conservation Society I have the honour of supporting thousands of incredible volunteers across Scotland tackling plastic litter on their local beaches through our Beachwatch Project. From the litter surveys they do, plastic is always at the top of the list and the passion and enthusiasm these volunteers have to tackle the issue inspires me every day to do the same.

What's the worst example of plastic pollution you've seen?

From bottles to wet wipes and nurdles to balloons, I have seen it all on beaches across Scotland. Since it only takes one item to potentially end the life of a magnificent creature like a leatherback turtle, each example can be heart-breaking. Some beaches feel spongy underfoot with layers of ropes and nets, some are covered in nurdles – the tiny pre-production plastic pellets – and some have wet wipes making up the strandline instead of seaweed.

What's the best solution to reducing plastic you've come across?

There is no one easy solution to reducing plastic in our seas; it has taken and will continue to take a joint effort by everyone including the public, industry and government to work together. A fantastic example of this is the 5p carrier bag charges – thanks to joint working by multiple organizations, public support and data, including evidence from our volunteers' Beachwatch surveys, we now have 5p carrier bag charges across the UK. Within one year we saw a 40 per cent reduction of carrier bags on our beaches, which shows the power of data and joint working.

Are there any changes you've made in your life to reduce plastic?

From the Plastic Challenge I have attempted each year I now use bamboo toothbrushes instead of plastic, shampoo bars instead of bottles and deodorant bars instead of plastic roll-ons, as well as carrying with me my MCS KeepCup, a stainless-steel water bottle, multiple fold-away bags and – my new favourite – a folding steel cutlery set! What I am incredibly proud of is to see my family and friends also taking on the challenge and in some cases I am running to catch up with their successes.

What makes you most annoyed when it comes to plastic?

Hidden plastics have to be one of the most annoying and frustrating things for me! When you are trying your best to reduce the amount of single-use plastic you use and you buy something only to find out within that cardboard box is a plastic bag or that book you ordered comes wrapped in cellophane. It can be a steep learning curve in reducing the amount of plastic you use but it also shows the importance of producers and retailers joining us on that journey and getting rid of unnecessary plastic wherever they can.

Do you have any top tips for getting rid of plastic?

Do your research and make some plastic-free friends! Social media has been a fantastic source of information for my plastic-reducing journey with shop and website recommendations as well as blogs on easy-to-swap-out items. The online plastic-free community is growing fast and I would encourage everyone to join it as it's much easier to take on the challenge with others who have tried and tested the path before you.

What do you think is the biggest challenge in getting rid of plastic?

I think the biggest challenge is to change how we design packaging and products to ensure nothing is ever wasted or thrown away. This will involve some big and brave changes from the plastics industry but I believe they are up to the challenge and the time to accept that challenge is now.

What do you think is the biggest opportunity in getting rid of plastic?

When I first started at MCS and my friends and family were describing my job to people they met they would say something along the lines of 'She works on something to do with fish'. Now, however, I hear everyone talking about 'marine litter' and 'ocean plastics' and when I speak to friends they tell *me* the latest piece of plastic-reducing news, which is fantastic! This, I believe, is the biggest opportunity – the world knows the problem and it is calling for change – now is the time for our world leaders to listen and take action.

What's the most impressive effort to reduce plastic you've seen – from an individual or a company?

I have met so many inspirational individuals, communities and organizations tackling the issue of ocean plastics and trying to reduce it. As it is currently the Year of Young People here in Scotland I would love to give a special shout-out to the Sunnyside Primary Ocean Defenders. These amazing ten- to eleven-year-olds have led the movement to reduce the amount of single-use plastic straws here with their 'Nae Straw at Aw' campaign. They have inspired villages, councils and even members of the Scottish Parliament to say 'Naw to a Straw' if they don't need it – true sea champions and an inspiration to all of us.

Starting Your Own Campaign

Of course, cleaning beaches or joining local groups may not be enough to contain your campaigning fervour, and why should it – there are so many opportunities for good campaigns to be run in your community that if you have the energy then nothing should stop you. If you are keen to run campaigns based around any of the products mentioned in this book or about a policy issue facing your area then here are some tools to help you get started. If you do have the energy but aren't sure where to start, here are a couple of campaign ideas to kick off with:

1. Bans on single-use plastic. As Tiza said in her interview earlier in the book, bans are simple and effective – they just work. Whether it's getting your local bar to get rid of straws or persuading your local government to ban the use of styrofoam food containers and plastic cutlery by all fast-food outlets in the area, think about the plastic you see on your streets and work out who can make the decision to stop it getting there.

2. More water fountains. Given how much of an issue plastic bottles are, one thing that every business and council can do is to install more water fountains (or at the very least make it clear they will offer free water to anyone asking).

Exercise 1

Write down here, in no more than fifty words, what the problem is that you're trying to change:

Write down here, in no more than fifty words, what change is required to solve the problem:

Every good campaign is made up of a series of steps that get someone, or several people, with power to use that power to make a change happen. Sometimes these steps can happen very quickly, sometimes they can take a long time – all of them started with just an idea. Now that you've outlined the problem and what you think the required change is you want to see, it's time to figure out who the person is you're directing it towards. Is it the CEO or sustainability officer of a business, a local politician or maybe your national representative? Every case will be different and, if in doubt, I recommend going for the most senior person you can find – the CEO, the chair of the board or head of local government. If it's not them then they'll soon tell you who is responsible.

Exercise 2

Write down the name of the person or people you are directing your campaign towards here:

You have your problem, your solution and your target. All that remains is a plan of action. As you think about what steps are needed to persuade your campaign target to act, it can be helpful to think in terms of a ladder, each rung representing an escalation. Even if you're very annoyed with your campaign target, try to remember that an ideal solution to your problem probably involves working together in the future, and so it's worth thinking about starting small and private in your tactics, before going big and public.

Here's one possible escalation route, although every situation will be different and you know your target and your community better than me. There are of course so many amazing, creative ways to campaign that this does not do them justice, but the more you do it and meet others and hear or read stories, the more ideas you'll get. It may also be the case that some rungs get repeated or grow in scale as time goes on; it may also be that after each escalation you meet with the campaign target to try and persuade them again. Please take this as beginners' guidance, not doctrine. For many local campaigns, however, this is a reasonably normal series of steps (and a number of Greenpeace campaigns have followed exactly the escalation laid out below to great effect). Each step is outlined in more detail in the following pages.

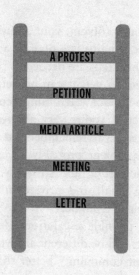

Every good campaign is made up of a series of steps that get people with power to use their power to make change happen.

Letter Writing

Across the world, writing letters or emails is one of the most effective ways at communicating directly with people in power. Whether it's to persuade a politician to vote a particular way or to persuade a business to stop stocking a particular product, your letter has every chance of being read and making a difference.

Introduction to Writing a Good Letter

Writing a good letter can be invaluable in persuading decision makers to agree to your campaign. What's more, a well-written letter in your own voice, tailored to the individual you are writing it to, can be far more effective than the thousands of identical emails they might get from an online campaign. Anyone can write a good campaign letter by following these five principles:

- Clear
- Short
- Personal
- Accurate
- Courteous

Clear

We all get carried away when we write or talk about something we care about. It is very easy to accidentally end up on a long rambling rant about an issue like plastic pollution. It is important to fight against this urge when you're drafting a campaign letter to a politician or business owner. Strip back your arguments to the bare essentials, because the most important thing is that the recipient of your letter understands exactly what you're asking, why you're asking it and what they need to do in response.

Before starting your letter, write out in a single sentence your answer to each of these three points.

1. What are you asking?
Is it for them to reduce their plastic footprint or to vote a particular way in parliament? Are you asking them to stop

stocking a particular product or to support a local clean-up activity? You might have more than one 'ask', in which case list more than one, but try not to overdo it – better to focus in on no more than a couple. This initial 'ask' should be in the first paragraph of your letter, making sure that whoever reads it knows right from the outset what you're writing for.

2. Why are you asking for this?

This is the heart of the letter, where you can bring in case studies, personal anecdotes and any other arguments as to why you're asking them to reduce plastic pollution one way or another. As well as keeping this part relatively short, it's important for any examples or arguments that you use to very clearly back up your 'ask'. For example, imagine you are writing a letter to a politician to persuade them to vote for a ban on plastic cutlery. One of the most commonly cited facts about plastic is that roughly a rubbish truck every minute enters the ocean. This might be true, and it is certainly a powerful statistic. However, a good letter should try to be as relevant as you can make it, so why not find a more specific fact about the amount of plastic cutlery entering our oceans or a case study of a country that has successfully banned plastic cutlery.

3. What would you like to see in your reader's response?

This could be naming the date of the vote you wish them to vote in and the way you want them to vote. You could finish with your vision of their cafe as a plastic-free zone, in which plastic stirrers and plastic-lined disposable coffee cups are a thing of the past. Whatever it is, keep a positive tone, be very clear and focus on how your desired response would be helping to move towards a world where plastic pollution is no longer choking our oceans.

Short

The person you are writing to, as someone who has decision-making power, is likely to be quite busy – or at the very least think of themselves as a busy person. By keeping your letter short, not only are they more likely to read it, but they will also be grateful they haven't had to trawl through an essay to get to the point. What do I mean by short? If you're sending your letter by post, then a side of A4 is easily long enough to make the necessary points.

It is tempting to include all the facts and statistics you know to make your case in the letter, but this would probably drown out your 'ask' and leave your reader confused. Try to use only one fact or statistic per point that you want to make, and make sure you pick the one that backs up your case most effectively or plays to the interests of the person you are writing to. It can be tempting to try to summarize a whole report or newspaper article that you found particularly powerful, but this is using up valuable space in your letter. To keep your letter short, better to just make reference to it and attach a copy (or a link to it) for them to read at their leisure.

Personal

Facts and statistics are essential to building a solid case, and they are often what grabs us in newspaper headlines or on social media. But it's important not to rely exclusively on numbers and distant examples, and to make sure to include at least a little bit on who you are and why this issue matters to you. Are you a parent at Christmas, annoyed at the amount of single-use plastic involved in gift giving and wrapping? Draw on that experience in your letter and you're helping your recipient relate to you as a person.

The most effective method for persuading anyone is to tell them a story from the heart.

We are far more likely to take advice from family, friends and people we know than we are from an article or television programme, so help your reader get to know you a bit. If you're a regular shopper at the supermarket you're writing to, make sure to tell them that in your letter – they're going to be more interested in the views of a good customer. Did you see something first-hand that pulled at the heartstrings: a coot or moorhen making its nest using bits of plastic or your favourite beach littered with drinks bottles? It might seem trivial to you, but these experiences can be just as effective in building your case as they help your reader understand where you're coming from. And, who knows, they may have had a similar experience to relate to.

Accurate

After you've finished writing, it's time to do a quick proofread. Simple grammar and spelling mistakes can really undermine the letter (get a friend to check, if you're not sure). If you've made any claims, make sure you can back them up if questioned, and if necessary, it's normally better to err on the safe side. For example, instead of saying '12.7 million tonnes of plastic enter the ocean every year', it is much better to say 'up to 12.7 million tonnes of plastic enter the ocean every year'. Any exaggeration could easily undermine your credibility and make your reader less likely to trust what you are asking them to do. You don't need to reference every claim you make in the letter, but you do need to be confident that if asked about anything you have said, you can easily find the source to substantiate your claim.

Courteous

Finally, remember to keep it polite and not angry or antagonistic. Would you respond better to a letter which aggravated you or one which encouraged you? Avoid the temptation to have a dig or make a snide remark, and be sure to address the recipient with the courtesy they would expect to receive and they will be far more likely to read on.

And that's it – if you follow these five principles, then chances are you've written an effective campaign letter. Time to click 'send' or lick the stamp and post it. If you would like more tips, then see the letter on the next page.

Dear Ms Wan,

What are you asking?

I am writing to you as the councillor responsible for my area about the need for more water fountains in the park on my street. Over the past few years I've noticed when I go for a run in the park that the amount of plastic rubbish being blown around has increased significantly. Although there are some bins, after a sunny day they are often overflowing and the plastic falls out the top, littering our area. This is something that more water fountains would help prevent.

Why are you writing?

One of the most common kinds of plastic I see in the park are plastic bottles. Plastic bottles and bottle caps are also some of the most commonly found plastic items on our beaches and in our oceans. Even the plastic in our park could end up in the ocean by getting blown into a waterway where it might float out to sea. A plastic bottle can take over four hundred years to break down once it is in the environment. Across the world the problem of plastic pollution is getting bigger, and just one of the biggest soft drinks

Interesting statistic to draw them in

companies is producing 120 billion bottles per year, and a million plastic bottles are bought around the world every minute. I can't see how this is sustainable if we want to stop so much plastic going into the ocean.

The most effective recognized solution to reducing plastic pollution is to reduce the amount we are producing. When it comes to plastic bottles, this is quite an easy job as people can carry around a reusable bottle instead. If we were to have water fountains in the park, it would encourage more people to carry around their own bottle or remove the need for them to go and buy bottled water as they could drink directly from the fountain. The Zoological Society of London and Selfridges are two examples of major venues that have managed to completely stop selling water in plastic bottles by installing more water fountains. In a report on plastic bottles by the Environmental Audit Committee, they found that installing water fountains could lead to a 65 per cent reduction in plastic water bottles — this kind of reduction would have a huge impact on

Examples to build your case

Evidence that directly backs up your request

the amount of plastic bottles we see in the park.

Making it personal

I love the park, and I find it upsetting to see plastic floating in the pond and tangled up in the hedges — it makes it look like our community doesn't care about where we live to leave it so messy and I feel embarrassed when friends and family come to visit. At the same time I also end up having to buy things like plastic bottles sometimes because that's the only option available. I,

What you would like to see in the reader's response

and I'm sure many others, would be extremely grateful if you would look to installing water fountains in the park as soon as possible. I would be happy to meet with you to discuss this further. Please write to me at the return address or you can ring me on XXXXXXXXX.

Yours sincerely,

Will McCallum

Meeting

You may want to write a letter first and then go to a meeting, or you might want to dive straight into seeing your target face to face if they're someone you already know. Whichever way round you do it, I can't emphasize enough how important speaking to someone directly can be. I can't count the number of times when despite months or even years of campaigning I've seen relatively little movement on an issue, but then after a good face-to-face meeting where we have a healthy discussion suddenly a company or politician decides to start taking action. Seeing someone in person can be your most effective tool at persuading them to act, and although it's here right at the start of your campaign tactics, it can be repeated at any stage.

What's the purpose of a meeting if all the information can be conveyed in writing? Put simply, it's you and your story that's the difference. When we come face to face with someone it's not just the words we use, but everything in the way we act in front of them that can help persuade. They can see for themselves that in most ways you're probably no different from them – just an ordinary person who cares about something enough to make a fuss, and who has something sensible to say on the matter.

How to Prepare for a Meeting

The best way to prepare for a meeting is much the same as writing a letter. Write down your main points: keep them short, clear, personal and accurate, and if you're the kind of person who gets nervous then practise saying them a few times. Write down a couple of the key facts you're planning to use – remember to make them as relevant as possible to the person you're meeting. If you can, try to get a friend or colleague to attend the meeting with you. If they're up for speaking, decide which points you're each going to say, but even if they're not then having someone there for solidarity is still a good idea.

Think about what you're going to take to the meeting. Have you seen a report that makes a lot of the arguments you want them to hear? If so, then take a copy with you. Are you trying to persuade them to sell reusable coffee cups so they don't have to rely on a single-use product in their premises? Take an example or two of the kind of cup they could sell. Have you come across any of their products on your beach or in your park? Make sure you have pictures you can show them. The more tools you have to help make your case, the more likely you'll be successful.

How to Conduct a Meeting

The most important thing in how you approach any meeting is that your ultimate aim is to walk out the door with a new ally – you're there to take this person and the organization they represent on a journey. Although you might be angry with something they've done or said, or frustrated that they can't see your point of view, at least in this first meeting it's important to keep things civil. If you managed to persuade someone to come with you, ask if they can take notes throughout the meeting. It can be hard after a meeting to remember everything that was said and agreed, so by having a note-taker you can help aid your memory later on.

It's likely you'll be asked to say your piece first in the meeting, in which case, as you've practised, go through the main points you want to make reasonably quickly in order to leave enough time for discussion afterwards. You don't need to use everything you have just yet – try to keep something in the back pocket for the ensuing conversation. Make sure to start and finish by outlining exactly what it is you're asking them to do. Once you've finished, be sure to ask them if they have any questions or want to focus on any particular point of what you've said, and ask what they think about your proposal.

As the discussion flows, remember to keep things clear and to the point, and if you catch yourself rambling then finish off with a question to put the focus back on them (it can help to wear a watch so you can keep an eye on the time). Remember that although face-to-face meetings can be a fantastic tool to persuade someone, they may not be in a position to make a decision right away. Being too pushy can be off-putting and if they're hesitant to make any commitments after you've asked a couple of times then it would be better at this stage to simply present a few more of the key facts and call an end to the discussion, promising to follow up later. If there are any questions that you can't answer, write a note and promise to get back to them with the answers.

After you've made your points, listened and responded to any queries and in general got to know your campaign target on a personal level, it's probably time to wrap up. As you bring the meeting to a close, make sure to go back over anything that either of you agreed to do, and promise to follow up with an email or letter to reiterate the agreed action. In your follow-up remind them of the discussion you had, answer any outstanding questions, reiterate your campaign 'ask' and give them a deadline to respond.

These things may all sound simple – and that's because they are. Being an effective lobbyist is as much about being organized and straight to the point as it is about being eloquent in the way you make your case. Of course, it may well be that you still haven't won your campaign, and don't worry, that would be normal. Lots of campaigns take several meetings and escalating tactics before they come to a resolution, but by following these meeting principles you've definitely got off to a good start.

The most important thing in how you approach any meeting is that your ultimate aim is to walk out the door with a new ally.

Using the Media

As a local person running a campaign with a local target, you are now of interest to your local media. If there's one thing that journalists are desperate for, it is authentic stories about the community they work in. Amidst the various advertisements and invitations to boring events, they're keen to find out what's making the people who read or watch their media tick. Congratulations – you and your campaign are now a story they want to cover. And if they don't know they want to cover it yet, then it's your job to persuade them.

It's easy to get intimidated doing media work, but try to remember that it's these journalists' job to find stories, and what you're doing is actually helpful. So long as you don't flood them with boring information, and remember that their job is also to find new stories so don't send them the same old story over and over again. It is important to contact them only when you have something new to say, and in this way you should hopefully build up a good relationship over a period of time where they trust the information you're giving them.

How to Get Their Attention

The first thing is to write a press release – a short summary of the news you're giving them, why it matters and who to contact. Make sure to write your release in the body of the email rather than as an attachment as you never know whether the attachment might result in your email getting caught by a spam filter. Give your press release a title that instantly tells them your news, and use the title in your email subject heading. Perhaps you've figured out exactly how many coffee cups per year your local council uses, in which case your opening sentence could read: 'Local campaign reveals council throws away 100,000 coffee cups per year'; or maybe you've handed in a petition: 'Locals call on council to change its ways and help end ocean plastics'. This opening sentence is what's known as your 'hook' – the thing that draws your reader in – and you need to make it as exciting as possible.

The best press releases are those that move swiftly

on to highlighting the key facts relating to the issue and the news as quickly as possible, so unless you're confident in your ability to write arresting prose, I recommend you stick to bullet points. In no more than five bullet points give your journalist the best statistics and a very brief summary of what has happened. If you're sending the press release about an upcoming event you hope they will cover, include the key details in these bullet points, as well as again at the end of the release. After these bullet points you can move on to a more narrative description of what the news is, why it is important and what you're asking to happen next. Make sure to avoid jargon or complicated language, and explain the issue and why it's relevant in the simplest possible terms. It's likely a journalist will be brand new to the issue and have less than a couple of minutes to digest what you've written, so you have to keep it clear and concise.

Vital to any press release is the quote from yourself or your nominated spokesperson. Make sure to include their name, any affiliation and a quotation that summarizes your campaign in a personal voice – this is the part of your press release that you want to be reproduced exactly as you've written it. At the bottom of the press release remember to include your contact details or the details of whoever you would like the journalist to get in touch with in case of questions.

Finally, nothing grabs attention better than some

good images. Pictures to match your press release will really help your story travel; even pictures taken on your phone can do the trick. If you have any accompanying photos, rather than attach them to your email and risk it getting sent into their spam folder, better to create an online album on a site like Flickr, and include a link at the bottom of your release.

Read it over at least twice before clicking 'send' and, if you can, get a friend to read it over too. Remember journalists are writers and simple things like spelling and grammar mistakes might get on their nerves. Make sure you've put the headlines in bold and the email looks well structured with any subheadings or bullet points properly formatted. After a final check, you're ready to send. Try to send it early in the morning so that you've got the whole day to follow up. If your local media outlet only goes to print each week, then make sure to send it about forty-eight hours before it comes out so that it has a chance of getting in the next issue whilst the news is still fresh.

After you've sent it to any radio, TV or newspaper addresses you could get hold of then it's time to pick up the phone to do a quick ring round. By this point in your campaign you might already have the number of a friendly journalist that got in touch – if this is the case, then give them a ring first. If you've got something newsworthy to say, they'll be grateful you came to them. If you don't already have any numbers, search online or in the pages of the local paper for the news desk number. Before you dial them, quickly rehearse how you're going to sell your story when they pick up the phone – have your press release in front of you. If they're interested, they'll probably ask you to resend your release, so it's best to do all this in front of a computer.

Here's an example of the press release that launched Greenpeace's plastic bottle campaign:

Greenpeace report reveals plastic footprint of world's largest soft drinks companies

Impactful exciting headline

Greenpeace UK has conducted the first ever comprehensive survey of the plastic footprints and policies of the top six global soft drinks brands: Coca-Cola, PepsiCo, Suntory, Danone, Dr Pepper Snapple and Nestlé.

Quick overview of the story

Despite plastic bottles forming a major source of ocean plastic pollution, the survey results reveal a woeful lack of action by the soft drinks industry to prevent their plastic bottles ending up in our oceans.

'The results are jaw-dropping,' said **Louise Edge, senior oceans campaigner at Greenpeace UK**. 'It's clear that if we're going to protect our oceans we need to end the age of throwaway plastic. These companies need to take drastic action now.'

Gripping quote that you want newspapers to print

Key findings:

Key points to the story including interesting facts highlighted

- Of the six companies surveyed, five sell a combined total of over **two million tonnes of plastic bottles each year – the same weight as over 10,000 blue whales**.
  - The largest brand **Coca-Cola refused to disclose**

Best statistic from any new information right at the top

the size of its plastic footprint**, making the actual total figure much higher.

  ○ When combined with plastic packaging used by the companies, the total figure rises to **a startling 3.6 million tonnes** a year (still excluding Coca-Cola).

- **The six companies use a combined average of just 6.6 per cent recycled plastic** in their bottles, despite producing fully recyclable bottles and placing the responsibility on their customers to recycle.

- **None of the companies surveyed have commitments, targets or timelines** to reduce the amount of single-use plastic bottles they use.

- A third of the companies surveyed currently have **no global targets to increase their use of recycled content** in their plastic bottles, and **none are aiming for 100 per cent recycled content** in an ambitious timeframe.

- Four out of the top six companies surveyed **do not consider the impact of plastic bottles on oceans in their product design** and development processes.

- Over the last ten years, the soft drinks industry has been **consistently decreasing their use of refillable bottles**, instead switching to yet more single-use plastic.

- Two-thirds of the soft drinks companies surveyed have a **global policy opposing the introduction of deposit return schemes** on drinks containers, which have boosted recycling and collection rates to over 80 per cent across the world, and more than 98 per cent in Germany.

Don't worry if your release doesn't have this many statistics; this is to give you an idea of how to present your facts

Louise Edge, senior oceans campaigner at Greenpeace UK, said:

'Our lives are awash with throwaway plastic. 12 million tonnes of the stuff is ending up in our oceans every year, where it harms marine life, spreads toxic chemicals and can take centuries to break down. We know that plastic bottles are a huge ocean-polluter and in the UK alone we dump 16 million of them in our environment every day.

'So it's not good enough for the biggest soft drinks companies in the world to pump out millions of tonnes of throwaway bottles and then blame everyone but themselves for their environmental impact. The results of this report are jaw-dropping. It's clear that if we're going to protect our oceans we need to end the age of throwaway plastic. These companies need to take drastic action now: phase out single-use plastic, embrace reusable packaging and make sure the remainder is made from 100 per cent recycled content.'

ENDS

Notes to editors

- See here for the full report, *Bottling It: the failure of major soft drinks companies to address ocean plastic pollution*: http://www.greenpeace.org.uk/sites/files/gpuk/Bottling-IT_FINAL.pdf
- Images of ocean plastic pollution can be found here (free to use with credit, requires registration): http://media.greenpeace.org/collection/27MZIFJJAYYJJ
- Video available on request.

For further information, interviews and comments, contact: Luke Massey

It is so satisfying to see a story you created in print or to hear it discussed on the radio, and media work can be some of the most rewarding campaigning you'll ever do. If you don't get the headline you're dreaming of though, there's no need to feel disappointed – it happens to everyone no matter how big the campaign. Journalists are busy people trying to sift out the most important stories of the day, and it may be you just caught them on a day when a few more things were happening. This is why social media has become so valuable, so that on those busy news days we all still have a means to get our story out to those that need to read it (you can even tweet the same journalists, letting them know you have a press release for them, if you get no response at all). If you still aren't getting much response, then think about writing a letter to your local paper – have a read of the letters page to see what sort of length and tone the letters they usually print are. Writing public letters like this can be a brilliant way to raise awareness amongst journalists at the newspaper.

Creating a Petition

Meeting your target, writing to them or even using the media to shine a spotlight on your campaign have still not worked? It's time to build the pressure by involving more people.

Starting petitions has never been easier. There are so many online campaigning organizations trying to help you on your way. Whether it's Change.org, Avaaz or 38 Degrees, the online tools to create a petition and share it widely are at your fingertips. Go to any of these websites and be guided through the process of setting up your own petition – you've got your target and the issue already, so it should be a breeze to set up.

Collecting Signatures

Once you've set it up, don't get disheartened if your petition doesn't suddenly fly with thousands of sign-ups a minute – it's perfectly normal. Make sure you're doing what you can to share it though: post it on your social media channels and email your friends asking them to sign. Make sure to ask those signing if they'll also share it with their friends – this way you can start to see the ripples get bigger. Have a look to see if any of your friends or anyone you have a connection to has got lots of online followers, then put in a special request to them to share it. Getting people who are already influential online to share things can really help get more people to sign.

See how it goes for the first couple of weeks, and set yourself the target of at least 200 signatures – enough to make most local politicians and businesses take notice; also a realistic number to get if you send it to everyone you know in the area and maybe put up a few signs with the link to the petition near the premises of your target or in an area affected by plastic pollution, to give it a bump.

Handing in Your Petition

Finally comes the hand-in moment. Maybe you've reached your target or maybe there's a vote coming up in government, but for whatever reason you've decided now is the time to hand in your petition. Have a think about how you want to present your petition. Do you want a prop, for example? Perhaps a paper straw per person that signed, a certificate you knock up in half an hour or maybe a large cardboard print-out with the total number of signatories – choose something that seems right for the occasion and looks nice in a photo. Perhaps you'll do it with other signatories or the people who have been helping you collect signatures, or maybe it will be a more informal affair where you'll just meet for coffee and tell your target the total number of people that signed.

By this stage in the ladder I'm sure you're in regular contact with your campaign target and know what the mood is, so contact them to let them know the day you'd like to hand it in. This way they can either agree to receive it themselves or send someone in their place (although I've been involved in plenty of deliveries that we had to shove through a letter box when we'd been refused an audience). If you

199

think it's appropriate, then it is probably worth letting the local media know you plan to hand it in, particularly if they've already written something about the campaign. If they don't come to take pictures, make sure you arrange for someone to take a few snaps to share with the people who signed the petition and on your social media channels.

Once you've handed it in, if you didn't get a meeting on the day it's pretty likely you'll get one now. If not, then it's time to step up a rung.

Organizing a Protest

When you think of a protest you probably conjure up images of angry-looking people waving placards – they may even all be dressed in vintage sixties paraphernalia. These kinds of protests do have their place and have helped achieve a huge amount, but are not really the kind I'm talking about. They would probably come on the next rung of the ladder, which is a step further than I'm going to go in this book – although if you are interested in mass organizing strategies then there are plenty of materials out there to read about the theory behind organizing this kind of protest, as well as guides on the more practical elements.

What I mean when I say organizing a protest is organizing a clever bit of direct communication with your target. Something that grabs their attention and makes them sit up and listen. It would be impossible to list even a fraction of the examples out there, but here are some thoughts on how to take things up a level:

1. **Snap, share and shame:** One of the simplest ways to draw attention to your campaign and pressure your target to act is to publicize what it is they're doing wrong. If your campaign is trying to persuade them to ditch straws then encourage everyone that signed your petition to take pictures every time they see a straw littering the pavement or on the floor of a bar. Tag the company you're targeting in your social media posts with the picture, asking them why they are still not acting. No company or politician wants to be associated with negative images, and this can be a quick way to help them realize that at the moment they're part of the problem, but you're giving them a chance to be a part of the solution.

2. **Craftivism:** Ask your friends and people that signed the petition to make something to send to your campaign target. This could be anything from origami fish to little glass jars filled with plastic you collect from the beach. Everyone's a sucker for a beautiful nick-nack and although they may not want to receive hundreds of them at once, it's a great tactic to push them to take action in a light-hearted way.

3. **Online activism:** Get all your petition signatories to start tweeting at the company director or the local

councillor. Create some sample tweets that people can use or direct people towards their Facebook page to leave comments asking them to act now. A few well-targeted social media posts can really make people feel the pressure. You could even take this a step further and organize with your fellow campaigners to all ring their customer service telephone number at the same time, inundating them with complaints.

4. **Leave it at the till:** Lots of companies just aren't listening to our complaints that they are over-packaging their products. One way to make them listen is to leave any plastic packaging you don't want at the till when you pay. You shouldn't have to feel responsible for producing so much waste and so next time you just can't bear to take it all home, carefully unwrap it and politely leave it in the shop you bought it in. If the cashier is upset then be sure to apologize – it's not their fault either.

5. **Return to sender:** If you're too shy to leave it at the till (don't worry, many of us are), then post it back to the companies that made the product. Next time Amazon or your supermarket deliver you their goods in mountains of plastic, pack it all up and send it back. If you find stray products containing microbeads or have an old pack of cotton buds you don't want to use, send them back to whatever company made them with a note about why you're sending them and saying that you would like to receive details about how they have recycled them in response.

From letter writing to protests, your campaign is a chance to be creative and get people involved in making a difference in your area. It should be fun – most people that go into campaigning enjoy what they're doing, because it feels good to be active and organizing. It's also tiring work and when things aren't going your way it can be dispiriting, so it's important to start small, set yourself achievable campaign goals to begin with and then work up to something bigger. Start with one cafe or one restaurant, before moving on to your whole area; start with influencing one business to get rid of plastic cutlery in their canteen before going after your council to ban all plastic cutlery.

There is no strict science to campaigning – it is an art and there is no better way to learn than on the job. As you gain experience and start to learn how best to persuade different people, businesses and politicians, you'll think of new tactics and new ways to get the attention you need to win. And when you win, make sure to take the time to celebrate. Each victory on the route to giving up plastic is worth a celebration and in the long term it's essential to make the time to reflect on the change you and your fellow campaigners have achieved.

There is no strict science to campaigning – it is an art and there is no better way to learn than on the job.

12

WHAT DOES THE FUTURE HOLD?

How to give up plastic – that was the promise of this book, a guide to giving up this material that is defining our generation; how to wean ourselves off something so commonplace. The answer is by working together. Alone we can play our part, and our actions are important, but it's when we are united that our efforts really make a difference. Like all threats facing our world today – some of which are even bigger and scarier than plastics – the answer does not lie in hiding behind closed doors, hoping the world will fix itself whilst we wait anxiously, twitching the curtains, hoping for the solution to be delivered to our doorstep. The answer to how to give up plastic rests in our ability to come together and call for ambitious action now – in our own lives, from companies and from our representatives in government.

There isn't a single path to giving up plastic and the routes will vary across countries and communities, but there *is* a single message: that we need to stop producing so much of it. Our throwaway culture has gone too far and the silver lining to the plastic bags strewn across our beaches is that it is forcing us to wake up to the fact that we cannot continue living in a society where it is acceptable to use something once before putting it in the bin with no thought as to the end of its life. The plastic littering our neighbourhoods is waking us up to the need to snap out of an economic model based on producing and consuming cheaper and cheaper goods with no consideration of the long-term cost to the environment of what we use. *How to Give Up Plastic* is more than a guide to ridding your home of unwanted plastic waste, it's a guide to joining the growing movement around the globe who say that plastic symbolizes a world of the past that didn't care enough for the environment we depend on.

It is true that plastic has already reached into the farthest corners of the world and been found in the stomachs of ocean creatures that have never come into contact with humans before. It is worrying that plastic production is still increasing and no major multinational company has yet come up with a realistic plan to reduce their use of it. I was shocked to hear that some geologists see the discovery of plastic in layers of rock as the sign of a new geological age where humanity's footprint has become visible – they call this the Anthropocene. However, it is also the case that awareness of this issue is stirring up something big across the world, unleashing a wave of dissatisfaction that so many of those I have spoken to share about the way in which we live our lives; that being so dependent on products that are doing us no good in the long term is a source of unease. Whether it's wandering a tideline littered with plastic fragments on the beach you love, watching YouTube videos of animals getting freed from a plastic snare or worrying about what all this plastic is doing to your health – you're reading this book because you know that the cost of inaction is too great.

Plastic pollution and the speed with which it is increasing can – like so many environmental problems – be quite scary and overwhelming. There's no point in pretending otherwise. If we try to minimize the problem in our minds, all we'll do is tell ourselves a lie that it's not that big and then our actions to solve it won't match the scale of the problem. Instead we should embrace the monumental task ahead and start using the immense power we have in our own lives and our community to tackle it. By facing up to the reality of the world we live in we can approach our future boldly, confident that the power of millions of people around the world fighting for the same thing cannot help but have an impact. Equally, we shouldn't say plastic has no benefits. It's precisely because it does that it's been so successful and we've all enjoyed its benefits. Cheap and hygienic, it has improved the quality of life for millions. Let's accept that, like the last drink at a party, it was a good idea at the time, but it's turned out pretty terribly.

There isn't a single path to giving up plastic and the routes will vary across countries and communities, but there *is* a single message: that we need to stop producing so much of it.

With every new study coming out about the impact of plastic, we are learning more and more about the way this material we rely on is shaping the environment we depend on. In the coming years this research will be joined by studies that tell us more about whether it is harming our health and the full extent to which it is harming our oceans. As our understanding of how plastic shapes us and the world we live in grows, I am confident that so too will the desire to give it up for the sake of our health, our environment and future generations. Technology, and our ability to organize and communicate more widely than ever before, is allowing us to play a bigger role than history has ever afforded us previously in shaping our society. The tools in this book are about more than just giving up plastic, they're about using your power as a citizen, voter, consumer and member of your community. There is nothing complicated or radical in taking the reins and demanding change. Common sense and an authentic story about why you care will take you a long way when it comes to helping others understand the need to give up plastic.

As you start on this journey, here is a reminder of the guiding principles behind your choices at home, the campaigns you run and the advice you give to friends, family and colleagues:

- **Refuse** plastic wherever you can – say no to single-use plastics that have become all too common in our lives on the go.
- **Reduce** plastic in your home and your workplace – switch to longer-lasting materials and work out where you can avoid buying plastics.
- **Reuse** – get hold of the reusable essentials to any plastic-free life like a water bottle and a coffee cup.
- **Recycle** – always make sure you dispose of the remaining plastic in your house responsibly, recycling it whenever possible.

But above all of these, the principle that in the face of the journey ahead I think is the most important:

- **Use your voice.** Let your friends know, let the shops you go to know, let your colleagues and your local paper know. The movement to give up plastic relies on millions more joining it – and you are essential to building it.

Plastic is not simply going to disappear overnight, and it's certainly not going to go away without a fight. The movement to give up plastic is going to take a monumental effort on behalf of millions of people across the world just like you who care about the environment and want future generations to enjoy the same beautiful oceans that we have been lucky enough to enjoy until now. It's a movement already made up of billions of individual acts, the ripples of which are being felt across our blue planet right up into the tallest skyscrapers. It may seem like an impossible task, but if there's one thing the last three years have taught us, it's that the world is changing at an unprecedented pace, and tasks that once seemed impossible are now within our grasp. At a time when stories of hope can seem in short supply, the movement to give up plastic is bringing together people from all backgrounds and all cultures and starting to paint a vision of a society that works together to create a better world for future generations.

The answer to how to give up plastic rests in our ability to come together and call for ambitious action now.

ACKNOWLEDGEMENTS

It should probably not have come as a surprise that writing a book is not a solo effort, but relies on the expertise, encouragement and ideas of many people – too large in number to mention all by name here. I am grateful to you all for the role you've played in inspiring these pages, and take full responsibility and apologize for any errors or mistakes they may contain and anyone I have failed to acknowledge.

With particular thanks to the incredible team at Greenpeace who campaign tirelessly for a better world, and the strength and commitment of those all over the world facing environmental challenges with humour, determination and anger every day. Particular thanks to those who have given their words and time to this book: Louise Edge, Luke Massey, Bonnie Wright, Willie Mackenzie, Tiza Mafira, Arifsyah Nasution, Afroz Shah, Catherine Gemmell, Amy Meek, Ella Meek, Tim Meek, Rachel McCallum, John Staniforth, Alice Ross, Angus McCallum, Jamie Szymkowiak, Ben Stewart, Emily Robertson (and the whole Penguin Life team), Alice Hunter, Grant Oakes, Sonny, and the Break Free From Plastic movement. And to those who have shared this plastic journey: Alexandra Sedgwick, Marcela Teran, Ariana Densham, Tisha Brown, Elena Polisano, Louisa Casson, Doug Parr, John Sauven, Emma Gibson, Pat Venditti, Damian Kahya, Dean Plant, Elisabeth Whitebread, Rosie Rogers, Paul Keenlyside, Rebecca Newsom, Fiona Nicholls, Frank Hewetson, Rachel Murray, Elsa Lee, Karen Rothwell, Sam Harding and Campaign to Protect Rural England, Fiona Llewellyn and the Zoological Society of London, Have You Got The Bottle, the Marine Conservation Society, eXXpedition, City to Sea, Environmental Investigation Agency, Fauna & Flora International, Break Free From Plastic, Kasia Nieduzak, Deborah McLean, Sebastian Seeney, Paul Morozzo, Tory Read, Melissa Shinn, Graham Forbes, John Hocevar, Paula Tejon Carbajal, Kate Melges, Sandra Schoettner, Manfred Santen, Christian Bussau, David Santillo, Paul Johnston, Melissa Wang, Eleanor Smith, Broken Spoke and so many more.

To friends and family whose support makes everything easier. And to Joe, who makes everything more fun.

A NOTE ON THE USE OF PLASTIC IN THIS BOOK

At Penguin Random House we have done everything we can to ensure that as little plastic as possible was used in the making of this book. For example, you might notice that there is no lamination on the front cover. However, despite all our efforts, we were unable to find a plastic-free glue which was strong enough to keep the book from falling apart, which just goes to show how entrenched our society's dependence on plastic currently is. We hope that books like this one represent an important step in thinking more deeply about the impact we all have on the planet.

#BREAKFREEFROMPLASTIC

#BreakFreeFromPlastic is a global movement envisioning a future free from plastic pollution. Since its launch in September 2016, over 1,060 groups from across the world have joined the movement to demand massive reductions in single-use plastics and to push for lasting solutions to the plastic pollution crisis. These organizations share the common values of environmental protection and social justice, which guide their work at the community level and represent a global, unified vision. Sign up at www.breakfreefromplastic.org.

GREENPEACE

Greenpeace is a global campaigning organization whose underlying goal is a green and peaceful world – an earth that is ecologically healthy and able to nurture life in all its diversity. We defend the natural world and promote peace by investigating, exposing and confronting environmental abuse and championing environmentally responsible and socially just solutions for our fragile environment. If you're interested in how you can join the movement, visit www.greenpeace.org.